AN APPRAISAL OF RATIONALISM

IN MODERN SCIENCE

BY

Patrick Johnson Mendie

Published by Lulu Publishers, USA

First Edition: 2017

ISBN: 978-1-365-69081-5

TABLE OF CONTENTS Pages

ACKNOWLEDGEMENTS

I wish to sincerely express my gratitude to God who is the giver of life and the rationale behind this work. My profound appreciation goes to my supervisor, Professor Andrew F. Uduigwomen for his thorough supervision and his inspiration towards the success of this research, his effort was indeed monumental. My appreciation goes to the Philosophy Department, University of Calabar for their mentorship which have strengthened my philosophical grounding.

Also I acknowledge the effort of my parent for their kind gesture in making my dreams comes true, through their moral and financial support. I must commend the contribution of Dr Uti Egbai and Dr Chris Akpan whose ideas have sharpened my understanding in philosophy of science, especially Philosophy of Particle Physics. I acknowledge my colleagues and classmates whose encouragement, challenge, criticism and company enforced my inspiration during this research.

Finally, i sincerely thank Brother Malachy Mendie, Isidore Mendie, Edidiong Mendie and Sister Rose Uwoh, Imaobong Essien and Dr. Donald Enu for their moral and financial support in my academic journey. I say a big thank you; may God almighty bless you richly for all your positive contributions. My appreciation also goes to non-academic staff of the Department of Philosophy for their kind support, i say thank you for all your support.

ABOUT THE BOOK

The beginning of modern science within 17th century gave birth to two schools of scientific thought, rationalism and empiricism. The operations of empiricism in science were defined as an experiential discipline; observation and all other faculty of sense data collections were the key factors of its greatest achievements. These empiricist doctrines became the determinant of what should be called scientific as shown in logical positivist's position. This understanding also brought into science a mechanistic understanding about the universe within Newtonian science. The rationalist on the other hand states that, it is rather through the aid of innate ideas or ratiocination that science can grow. This essay entitled "an appraisal of rationalism in modern science" therefore carried out a formal assessment of rationalism in modern science and argues that rationalism is the foundation of modern scientific knowledge, thus, rationalism has featured more in modern science than empiricism. The work contended that to make sense from the sensible world, science needs intelligibility that comes from the process of ratiocination, as this has played the leading role in most scientific investigations in modern science. To build support for this argument, the work examined some moments of rationality; moments of formation of relativity and quantum mechanics, thus some rationalists' philosophers of science whose works have strengthened the rationalist contents of science were critically examined. Given the nature of this work, we applied the philosophical method of critical analysis. The work further posited that, the picture of science as empirical discipline has lingered for too long, such that modern scientists continue to think that science is through sense dependence. The work argued that, this general perception of science as sense dependent has tilted away from empiricism and leaned more towards rationalism in the modern period.

DEDICATION

This book is dedicated to God almighty and to my late father, Deacon Johnson Mendie.

CHAPTER ONE

INTRODUCTION

Most human beings by nature are naturally faced with the drive to aptly investigate and to give meanings to the general composition of the universe. By this quest, if one is asked to explain the composition of the universe, some will be tempted to make a list of stars, moon, sun, oceans, rivers, buildings, and so on. These are few elements of reality that can be sensed directly through human sense data collections via sense organs and hence justified by experience. To the other, the general composition of the universe consists of unseen elements of reality which can only be understood through the process of rationalism. The challenge here is, of which method should scientists consider first during any scientific investigation of the universe? In other words, what should be the foundation for scientific research?

Within seventeenth and early eighteenth century's philosophical discourse the above questions brought in a heavy intellectual challenge, of what could be the most authentic means of understanding science. Because of this challenge, the methods of rationalism (deduction) and empiricism (induction) emerged in attempt to guide scientists during any scientific investigations to proffer solutions to all scientific problems. Some questions were developed within the philosophical discourse like: can innate ideas guide us to sustain a better place for science? Or should scientific method be guided by human experience alone?

However, understanding the mechanistic universe in the works of Isaac Newton in modern science, he explained the universe as a product of the five senses; sight, hearing, touching, smelling and tasting, and through this process scientific laws were developed making prediction to be possible. The outcome of Newtonian physics was a product of these perceptual understanding as the building blocks of science guided by reason. This is why Dennis Igwe agrees that, the laws of Newton were based on that understanding of the universe as a product of observation (351). The science of Isaac Newton introduced us to a modern approach of how the laws of the universe works, a universe that operates like a machine with laws to unravel the mystery behind nature.

A further reflection on the world-view of Newtonian science, which was the foundation of modern science, one could discover how empirical explanatory model became the driving force of scientific predictions and the handmaid of causality predominantly an outcome of induction and justified by fact deduced from laws and theories (Chalmers, 8), this development is believed by philosophers of science to be founded only by the principle of rationalism. Chalmers presented it to shows how scientific predictions were possible through observations, formation of laws and theories and later prediction becomes possible. This achievement in science became possible because scientific laws and theories can be deduced through the aids of rationalism. Because laws and theories themselves are not things one can observe, but are derived through mental construct used in solving scientific problems. This process according to Bryan Magee is the foundation of the heliocentric approach of the universe that replace the Ptolemaic cosmology by Nicholas Copernicus and also a product of what gave Isaac Newton the driving force to find certainty and absolutism in knowledge through laws and experiment guided by reason (68). The application of reason has produced enormous growth in science, and intelligibility is therefore believed to be the prominent factor that has possibly directed all scientific investigations.

Meanwhile, in 1905, Albert Einstein introduced the special theory of relativity that revolutionized classical science and turned science upon its modern course. This revolution lay in the understanding that guarantees that all the laws of science (physics) are the same in all inertial reference frames (Essien 223). In other words, in every inertial reference frame, all physical laws remain unchanged. Einstein proved the inevitability of rationalism in modern science by establishing a rational colouration in his special theory of relativity and ten year later the same was done to his general relativity of 1915. In special relativity, Einstein predicted that space and time are not two distinctive entities as adopted in Newtonian science, but one that cannot be separated and general relativity explains the rationality that exist in a gravitational field which are both derivable from the observers frame of reference (Gribanov 217). The predictions of Einstein relativity proved that space and time exist as one and the same thing called space-time. This is possible because, space and time is seen as a property of matter such that nothing could be conceived outside the arena of space-time continuum (Essien 223). Stephen Hawking agrees with Einstein when he considers an event "as something that happens at a particular point in space and at a particular time" (Hawking 24). From time to time one cannot separate space from time, making it possible for Einstein to establish space-time in relativity. General relativity predicted that light bends in a gravitational field (Gibbs 469). This prediction shows that when large matter falls within an empty space, there is always a space bend which may cause curvature of light within the space line because of the bending space. All these experiments were possible through un-experimented facts derived through imagination called *thought experiment.* Thus, Einstein states,

> I was sitting on a chair in my patent office in Bern. Suddenly a thought struck me: If a man falls freely, he would not feel his weight. I was taken back. This simple thought experiment made a deep impression on me. This led me to the theory of gravity (201)

This thought experiment became the foundation of Einstein contributions to science that only became possible through the process of ratiocination.

Another moment of rationalism this work shall discuss is the moment of development of quantum mechanics. Since the guiding principles of Einstein relativity and Newtonian physics could only solve the problems of the macro world a new science was needed to study the micro world (Mamadu 450). Through Marx Planck's action of "quanta" in 1900 a new science called quantum mechanics was developed to give meaning and explanation to sub-atomic particles of the micro world which was not accounted for by relativity and even by classical laws of

Isaac Newton (Egbai 14). Quantum mechanics pursued the aspect of rationalism in science paying much attention to what we cannot experience; by this it investigates the unobservable part of nature, the world of quarks, spins, leptons, photons, mesons, plasmas, electrons and protons, et cetera. The principles of indeterminacy and uncertainty, and the Schrödinger's cat experiment were all proven without the aid of physical experience but through thought experiment, because realities at the subatomic realm are unseen and in constant randomness and chaos only determined with probabilities (Akpan 75).

To show the strength of rationalism in science, the work further examines some rationalists philosophers of science, whose ideas contributed to the growth and development of science. The contributions of Isaac Newton, Karl Popper, Leibniz, Imre Lakatos were largely influenced by intellectual and imaginative originality, thus, a deeper study of modern science reveals a greater place of reason than experience in the study. In other words, to make sense from the sensible, science needs intelligibility that comes through the process of ratiocination. To further strengthen this position, the work examines moments of scientific development and revolution that have shown the rationalist contents of science, which includes moments of formation of relativity, and quantum mechanics as explained above. The impetus of this research is captured on the assessment of rationalism as the foundation of modern science; highlighting the moments of science that have produced a remarkable aspect in technological development which are all founded on the principle of rationality. The epistemological search for what could be the foundation of scientific knowledge in modern science has been one of the most challenging issues in philosophy of science. Most empiricists believe that through experiment, observation, and collection of sense data, an indubitable knowledge can be acquired. Most rationalists postulate that, it is rather through reason or innate ideas that science can arrive at indubitable knowledge about the entire mechanism of the universe. There is one sided perception that science is more of empirical discipline within the scientific world today.

The picture of science as empirical discipline has lingered for too long, such that modern scientists continue to think that their discipline is sense dependent. The work argues that, this general perception of science as sense dependent has tilted away from empiricism and leaned more towards rationalism. This is so because to understand the theory of relativity and quantum mechanics, sometimes regarded as the hallmark of modern science, scientists need intelligible power through the process of ratiocination.

The main objective of this work is to establish the prominent role played by rationalism in the birth and growth of modern science.

Other objectives are:

> To highlight the relevance of rationalism in modern science and its contribution to knowledge.
>
> To examine contributions from some rationalist philosophers whose works have strengthened the growth and development of modern science.
>
> To show the diminishing influence of empiricism in modern science (Theory of relativity and Quantum mechanics).

This work is significant in two perspectives, theoretically and practically. Theoretically, it is important in the investigation of the rationalist models of science and hence contributes to the growing literature in philosophy of science. Because of its thought-provoking nature, it is also relevant as it will encourage and provoke further research in the area of epistemology and philosophy of science. Practically, it will reawaken the consciousness of scientists and philosophers of science to the fact that science as a discipline is embracing rationalism in the modern era. This will lead to the revision in the existing literature in science and also will make philosophers of science essential participators in the growth and development of science. This research is justified because it attempt to participate in the debate between rationalism and empiricism in modern science which has become a topical issue in philosophy of science that needs urgent attention. Our position is that because of the enormous functions of innate ideas, thus, in order to solve the problems of the universe, scientists need intelligibility which can be derived through ratiocination.

Therefore, this research derives justification in its attempt to examine some philosophers of science whose ideas have shaped rationalism in science. The work is justified by producing vital information on credible moments in science that obey the laws of rationalism in science. This work is also justified because the problem of over dependent on empiricism in modern science is yet to be resolve.

This study employs the analytic, descriptive, logical argumentative and evaluative methods. The method of analysis is applied in the area of explication of the various concepts, the literature review and exposition of Isaac Newton's contribution to modern science.

The method of description and logical argumentation are applied in the discourse of the development of rationalism in modern science. The method of evaluation is adopted in the proposal of rationalism as the foundation of modern science.

This work falls under the area of philosophy of science, with much emphasis on rationalism in Modern Science. The scope of this work is limited to rationalism in modern science and the method of rationalism as embedded in the contributions of the rationalist philosophers. The work covers moments that have given science a new rationalist's look, (Moments of formation of relativity and quantum mechanics) which have shown to be more reliable in the growth and development of modern science. In this work, modern science is taken to mean `the era of theory of relativity and quantum mechanics.

The philosophical school of empiricism as used in this work is seen as one of the most unique schools in modern philosophy that contributions to knowledge acquisitions, it lends credence to the study of nature through the aids of the sense organs, the senses of sight, hearing, smelling, feeling and tasting. All these for the empiricists are essential means of getting authentic knowledge about the nature, scope and origin of the universe.

Empiricism as a philosophical school belief in sense perception, induction, and, that, there are no innate ideas in the formation of knowledge. G.O. Ozumba, in his thought provoking essay entitled "*isms in philosophy"* critically expatiated on the concept of empiricism as a school of thought that holds that true knowledge can be gotten through sense experience. The five senses of hearing, seeing, smelling, tasting and feeling are five important ways of getting acquainted with the external world (*A Concise Introduction to Philosophy and Logic*, 47). He further defines empiricism in his *Concise Introduction to Epistemology* (48) as an epistemological school that bases knowledge on experience, observation or experiment rather than theory. Empiricism holds that certainty in knowledge can solely be reached through empirical experience. Thus, the tenet of empiricism is directly opposed to rationalism. A. F Uduigwomen (*A Textbook of Philosophy and History of Science, 140,*) further elaborates that, empiricism can be described as an epistemological movement which holds that nothing around us can be known to be real unless its existence is revealed in or referable from information we gain directly through sense experience. Notable modern empiricists are John Locke, George Berkeley and David Hume. John Locke is regarded by some scholars as the father of modern empiricism because of his consistency in arguing for empiricism.

The work investigates Modern science as the centre of modern civilization, an era of rapid development in science and technology. Modern science brought in a more systematic and empirical justification of what science should look like as against the mythological and speculative approach to science by the ancient philosophers. It came in with a systematized step by step approach in carrying out effective analysis of all scientific evaluations. Thus, it was an era that enjoyed a more comprehensive outlook of the nature, scope and properties of the entire universe through ratiocination as the foundation. Thus, in the analysis of Uduigwomen, there are, at least five criteria of modern science which include:

I. Inter-subjectivity or objectivity
II. Precision, specificity and definiteness
III. Reliability
IV. Coherence or systematicity and
V. Comprehensiveness (19)

The above criteria illustrate the nature and the place of modern science in the development of a precise approach to knowledge and this produced a great advancement for a complementary knowledge about the universe in technological development, scientific innovations, new discoveries and a greater inspiration of new possibilities and a better approach to longstanding scientific issues.

Thus, science is as old as humanity. Science began in attempt to render a sustainable solution to the surrounding problems of human beings in the environment. The phrase "modern science" is used to describe an era of science where systematicity came into the study of the universe. The background to modern science can be traceable to the era of Copernicus, Beacon, Kepler and Galileo who first gave science a modern outlook and subjected science into a systematic approach through evaluation of fact, parting out from the mythological methods of understanding of nature. It therefore means that before now, science did not attain systematic approaches, mysticism, sorcery, voodooism, magic, witchcraft, diablerie, mojo, thaumaturgy, necromancy and many others were seen to be scientific. Thus, the idea of observation, evaluation and testing of facts came into science as a result of Galileo's effort in picturing science as a systematic discipline. This influences the positivists' notion of science and up till today has influenced the method of science. This idea of systematicity is what helps Isaac Newton to observe the stars and the falling apple that gave birth to the law of gravity and in the invention of his telescope; and since then science is seen as an empirical discipline that has given rise to the production of automobiles, industrial machines and many others. Science, today now deals with not only the palpable fact but it studies the unseen realities of the sub-macroscopic world of elementary particle physics under quantum mechanics and string theory. In this work, modern science is taken to mean 'the era of theory of relativity and quantum mechanics.

In understanding this research, Particle Physics is another key words. Particle physics began when scientists and philosophers limited their understanding of nature to tiny indivisible substance. In other words, elementary particles were believed to become the building blocks that made up the macro world of matter. In modern physics, the discipline which studies the underlining element of elementary particles that make up the physical universe is known as particle physics.

Particle physics studies theories and existence of atoms, protons, electrons, mesons, plasmas, leptons, quarks, etc. In other words, the principle surrounding the existence of all sub atomic particles, their nature, scope, and origin is holistically subjected under the area of particle physics.

Rationalism in this work renders the philosophical explanation on the place of reason in modern thinking. The idea of rationalism came into philosophical discourse, to portray a kind of belief system that takes credence of innate ideas as the root of all knowledge acquisition. It can also be seen as a philosophical school that posits that, authentic knowledge comes to us only through reason. And it is only through deductive ratiocination that one can get full grasp of the entire nature and scope of the universe.

A. F. Uduigwomen explains this view in his book entitled *A Textbook of Philosophy and History of Science,* when he asserts that, rationalism is the philosophical school that stresses the primacy of reason over any of the faculty in the acquisition of knowledge. He further states that for the rationalist, reason is the primary means of determining truth and of justifying knowledge, thus, ideas for the rationalist originate from the mind (139). G O. Ozumba agrees with Uduigwomen when he contends that, any knowledge that is based on sound reasoning, logic, mathematical procedures and other extra-sensory means is termed rational knowledge (*A Consise Introduction to Epistemology*, 59). The rationalists believe that the *a priori* supersedes the *a posteriori* in the acquisition of authentic knowledge. *A priori* knowledge is a knowledge that is independent of sense experience, meaning reason alone is capable of leading us to true knowledge without sense experience.

Notable philosophers of this tradition include, Rene Descartes, Spinoza and Leibniz.

Science is also used in this work to mean etymologically traced from the Latin root word "Scientia", in German "Wissenchaft', all of which refer to a systematic body of knowledge. This body of knowledge is derived through the study of the physical nature of the universe.

In the same understanding, A.F. Uduigwomen defines science as a knowledge arranged in an organised or orderly manner especially knowledge derived from experience, observation and experimentation" (20). This explains that science deals with what can be seen, touched, tasted, heard and smelt alone. A. F. Chalmers agreed with Uduigwomen when he asserted that, scientists see themselves as following the empirical method of physics, which for them consists of the collection of facts by means of careful observation and experiment and subsequent derivation of laws and theories (xvi).

This informs us that most scholars perceive the method of science as the handmaid of empiricism alone or as a sense dependent discipline. Following this thought system, Princewill Alozie in his *Philosophy of Physics* conceptualizes science as a discipline tasked to give explanation and understanding of the universe as well as fashion the relationship that exists between phenomena.

In this research, we define science as that which studies the material and non-material, the empirical and non-empirical aspect of the universe in a conglomeration of every aspect of the universe, man, non-humans, force, speed, energy, momentum, subatomic particles like quarks, leptons, and photons in a comprehensive approach.

Thought Experiment the last in this series is a kind of experiment in science and in philosophy that is carried out through the process of ratiocination. Here the mind undergoes certain active imagination of innate events without having any connection with experience to uncover hidden truth about reality which could not be possible with perception.

CHAPTER TWO
LITERATURE REVIEW

One of the fundamental problems within modern science was the dichotomy between empiricism and rationalism in science. Empiricism in science illustrates a tradition of belief that the mystery behind the existing universe of science can only be understood through sense data collection as the foundation of knowledge; as such, observation and experiment alone are the only criteria for all scientific achievements. Conversely, the rationalist, who also felt that the opposite is the case, took an opposite direction that fronted reason or innate ideas as what defined authentic knowledge. In this work, it becomes imperative to review our literature in these two thought systems to ascertain the various contributions of scholars on issues of rationalism and empiricism in modern science.

A. F. Uduigwomen in his book *A Textbook of History and Philosophy of Science;* gives us a vivid illustration of the nature and scope of empiricists' notion of science. Precisely, science is a kind of knowledge arranged in an organized or orderly manner, especially knowledge derived from experience, observation and experimentation (20). He further asserts that empiricists anchor their position on the fact that natural science scrupulously appeals to observation and experimentation in testing its theories (146).

From the definition of science above, it is obvious that the entire gamut of science is influenced by mere observation and experiment; observation by implication demands the analysis of reality from the way nature presents to us through the five senses. In the empiricist framework, innate knowledge is not admitted. Unfortunately strict abidance to the tenet of empiricism will lead to the elimination of the knowledge of the non-empirical realm of nature from the corpus of scientific knowledge. Since scientific knowledge in the empiricist scheme is reducible to the empirically observable, then one can question the authenticity of such knowledge since the observation of a phenomenon by two individuals may differ considerably. As such, to understand science, the role of reason is mandatory because most scientific progress starts from the theoretical before practical, and theories are created from the human ability to reason on the state of nature.

Chris O. Ijiomah in his *Harmonious Monism: A Philosophical Logic of Explanation for Ontological Issues in Supernaturalism in African Thought* caps the idea of science from empiricist tradition when he contends that every reality is composed of matter, which is the element of changeableness of things (71). From Ijiomah's background, it suggests that matter is the primary element of science with emphasis on entities that are purely materialistic. Matter as illustrated by Ijiomah presents to us the welcoming moment of materialism, a philosophical school that obeys the experiential world of matter as the only existing primordial force of scientific research. Ijiomah did not note the idea of dark matter which is not materialistic, but exist as part of the predictions of string theory. Therefore, the work does not picture reality holistically because any scientific discourse without reason makes it an incomplete approach to science.

John R. Reitz and Frederick J. Milford in their book entitled, *Foundation of Electromagnetic Theory,* also agree with Ijiomah's assertion that matter exists in all reality including electric and magnetic field. Thus, they contend that "it is our belief that a full understanding of the electric and magnetic field inside matter can be obtained only after the atomic nature of matter can be appreciated" (v). The agreement here is that nature is unfulfilled without a close look at the nature of matter in every reality. In other words, matter is also considered in the study of electromagnetic theory relating to electric currents within a magnetic field.

The question that we may ponder in the moment is: If reality only obeys the world of matter alone, and matter is all that science should investigate or scientific methods should only deal with, what could be the fate of the non material element of reality? is reason not necessary in the understanding of matter? This will give us a clear picture that materialism alone is not a complete way of studying the entire universe, to study the material world reason must guide our understanding of what we see, feel, hear or taste.

C. F. Chalmers is another scholar who has created much impact in his explanation of the method of empiricism in science. In his book entitled, *What is this Thing called Science,* he eloquently explicates on the method, scope and nature of Western science. In his discussion of the method of science he avers that a reflection on the worldview of the Newtonian physics or classical mechanics would reveal the insight that explanation and prediction, cause and effect (causality) constitute the building block of any scientific investigation. Through certain physical observation of the phenomenon, induction is made possible, which further develops or generates to laws and theories. Thus, deductions are established from the laws which end up in explanations and predictions as a mechanism for the science of that age (Chalmers, 8).

Chalmers highlighted three stages in which science moves in its aim of prediction.

1. Laws and theories
2. Initial conditions
3. Explanation and Predictions (10).

In view of the above stages, Chalmers only succeeded in presenting to us an explanatory model from the empiricist background that science begins through observation through which induction is made; it is induction that further gives rise to laws and theories. Thus, scientists make their deduction from the laws and further make prediction and explanation of natural occurrences. The information here is that induction gotten from observation drives the movement of science. In other words, the method of science is induction derived from sense data collection. Induction here has a major limitation in the study of science of which one of its major consequences is that sometimes it leads scientists into the fallacy of hasty generalization, if the samples analysed are gotten from few enumerations of cases. Chalmers at this juncture limited scientific method as fundamentally the handwork derived through observation and testing of palpable facts, this work argues here that for every observable to be known there is a greater place for reason than experience.

Mesembe I. Edet in his critical essay, *Being as Missing Links,* also agrees with Chalmers contention when he opines that modern scientific knowledge begins with empirical facts and is verifiable by other empirical facts. Thus, he stated that there are those philosophers of the empiricist or positivist bent who become so engrossed in the empirical method and the sensible beings of the world, that they make the sensible world the subject of their philosophy. He further contends that, for these philosophers it is the sensible, so to speak that has meaning and make sense and the spiritual or metaphysical is sophistry, meaningless, sheer nonsense, which needs to be discarded, or at least, it is obscure and questionable as a mode of being. Scientists in their subscription to the empiricist tradition hold that scientific knowledge is confined to the sensible or the empirically observable. In this vein, Asouzu explains that these scientists assume that only the sensible or experienceable is real. (*Journal of Complementary Reflection: Studies in Asouzu,* 28).

It is no more a doubting fact from the above purview that most scholars according to Mesembe imbibe the view that knowledge of science within the culture of experimental facts is the only criterion for science. The question is why should only sensible or experimental facts be seen as scientific? Can experience alone define science? This research, argues that such facts has a deteriorating approach that may not fasten scientific growth; thus, there is need for re-evaluation, because it lacks a holistic scientific methodology that lends credence to every aspect of reality especially from the perspective of non-experience.

Roberto Torretti in his book, *The Philosophy of Physics*, also gave us another vivid illustration of the magnificent impact of both rationalism and empiricism in modern science (Physics) when he makes a maximum exploitation of distinctive feature of Physics as a mathematical and experimental discipline. He emphasizes that experimentation naturally comes up in every aspect of practical art, namely, cooking, gardening, metallurgy (3). He further established the connection between the doctrine of experimentation and Aristotelian Philosophy. Aristotle and Galileo viewed experimentation as the sole source of knowledge. He points out that Galileo carried out experiment that informed him of the nature of the universe; through observation both agreed that a ship can float in a deep sea than a shallow one, this assertion was based on the experience that large object float better on deep sea. He posits further that Aristotle's philosophy of science considers matter as the primordial force of science; he gave us a theory of motion that avers that every terrestrial body had a natural motion towards the centre of the universe (Butterfield 15). As such Aristotle is seen by many as an empiricist philosopher. Empiricist in the sense that his idea agrees to the methods of sense data collection in science, but through a close examination, for every discovery of Aristotle there was an active reasoning power that supersedes experience.

Princewill Alozie in his *Philosophy of Physics,* caps observation as the preoccupation of scientists and further offers an elucidated explanation of what constitutes observation. For him observation includes the working of sense organs, this approach of sensory organs forms the found-spring of science. He further avers, that, through this medium the term empiricism or positivism is acquired (19). He further posited that observation is tied together with the functions of our senses of sight, smell, taste, touch, and hearing, which have all played an important role in the quest for making sense out of our physical environment (20). Alozie's view can be said to have agreed with the position adopted by Chalmers, because both ordinarily proffered the method of science as that which is gotten from observation and/or empirical justification. Alozie in his *Philosophy of Physics* further asserts that the major tool of science is geared toward explanation and understanding the universe through constant investigation of the relationship that exists between the phenomena (18). Phenomena in this sense deals with a kind of remarkable object of perception or a situation that is observed to exist or happen. Phenomena are objects of perception as it's appear to us and how it colours our holistic understanding of the physical world of matter. These phenomena will be nothing if reason is not fully applied to understand; as such it is reason that guides our understanding of the universe.

Carl F. Craver added another significant idea on the driving force of empiricism in science, when he also added more explanation on the aspect of phenomena. In the introductory section of his *Structure of Scientific Theories,* he gave a new insight of the world of phenomena in science as the building block of scientific theories. He maintains that, the central aim and objective of science is to develop theories that exhibit patterns in a domain of phenomena. These theories are used by scientists to control, describe, design, explain, investigate, explore, organize and predict the items in that domain (*Blackwell Guides to the Philosophy of science*, 55). As such, the world of phenomena becomes one of the most instrumental aspects of all scientific investigations. Craver further contends that, this method was the key and predominant guide to Logical Positivism. Thus, he states that, the central thesis of logical positivist philosophy of science is an analysis of theories from the domain of empirically interpreted deductive axiomatic system. A system that is founded on the justification of empirical palpable facts liken to their verifiability principle as the only criterion for scientific investigations (55). Craver's explanation about science is another work that enriches this research and deepens our understanding of how science develops, because it enables us to appraise rationalism in modern science.

Also, Jim Woodward in his critical essay entitled *Explanation* maintained that, the properties of the entire universe can easily be understudied using a Deductive-nomological model of Explanation. This model was first advocated by Carl Hempels as a sufficient and authentic way of giving an explicit account of the universe through a method known as Deductive-nomological model; a model of explanation that gives credence to observational factors as the base for deductive analysis. In empiricist doctrine, scientific deductions which are rooted to explanans are all captured through facts gotten from the explanandum; explanandum here are facts from observations of the physical properties of the universe; as such anything outside the terrain of observation the explanans is unjustified. Thus, Woodward contends that within the explanatory methods in Hempels model, the explanandum entail the explanans (Uduigwomen 299; Woodward 37).

From the above exposition by Woodward, all deductive-nomological model of explanation in science begin from inductions which are derived from observational factors (explanandum). This model of explanation, does not really give us a full model of understanding the universe in totality; this is because, we have identified it as not fully complete because certain aspect of reality do not obey the method of induction. In other words, it is not complete in the sense that it captures realities from phenomena (world of matter) as the explanandum. Thus, from this, Hempels deductive-nomological model of scientific research starts from observation rooted to the empirical description of the universe and neglecting the terrain of the unseen world of sub-atomic minute world of elementary particles of quantum mechanics which is today the power house of modern scientific developments, resulting in Laser inventions, Smart phones, Ipads, Computers, Magnetic Resonance Imaging (MRI) Scanners, Radars used in keeping track of airplanes, Global Positioning System (GPS) and other digital technologies presently helping man cope with the problems of his environment. The unseen reality do not obey induction as a method because nothing can be seen with human eye within the subatomic reality, thus, the place of reason is needed for scientific explanation.

Pagel, Heinz on *The Cosmic Code: Quantum Physics as the Language of Nature,* also argued concerning the idea of Newton, he states, that, the underlying determinism is the view that everything in the universe, including all the motions, from the smallest to the largest occurs in a way that can be predicted with absolute accuracy using the laws of Newton (4). Pagels' view gave us a background on the mechanistic laws of the universe within the terrain of Newtonian science, and era dominated by empiricism and justified by reason.

Herbert Butterfield in his work on *The Origin of Modern Science* captures the futility of empiricism in science, by illustrating a remarkable assertion in support of observation and experiment. He gave us a full description of modern science as purely from empiricists' background; concerning the idea of motion. He posits that, Aristotle first gave meaning to the concept of motion in science. Thus, Aristotelian teachings on motion were the handwork of Aristotle's observations and explanations of the way the entire universe operates. Aristotelian theory of motion states that, all heavy terrestrial bodies had a natural motion towards the centre of the universe (15). This is further explained on Aristotle notion of inertia, the idea that a body could continue to move only so long as a mover was actually in contact with it, imparting motion to it all the time, and once the mover ceased to operate, the movement stopped-the body fell straight to earth or dropped suddenly to rest. (*The Origin of Modern Science* 15-16).

Butterfield, from the above purview, gave us an intellectual illustration on the origin of modern science in relation to Aristotelian idea of motion. One can easily observe that the master ideas of Aristotelian science were influenced by observational knowledge, a clear agreement and evidence of empiricism, which influenced other scholars after

Aristotle in modern science. For example, the works of Galileo and Isaac Newton on the three laws of motion, the law of universal gravity were all influenced by Aristotle's doctrine of inertia, though with diverging explanatory approaches. However, observation is very crucial to both approaches. The question that may arise here is does motion exist only on physical object? Can unseen particles of nature participate in motion? The answer is 'yes', because, sub-atomic particles moves from one place to another in constant chaos with one another, but their movement are not absolutely determine, they cannot be predicted either, but always indeterminable, they can only be determined with uncertainty and justified by reason. In other words, since subatomic particles do not obey classical laws of science, it is impossible to conclude that the empiricists' idea on laws of motion can give us a definable holistic explanation of how objects or subatomic particles move. Another question one may ask here is, can we see motion? Motion is just a concept that is used to explain the movement of object through the process of ratiocination, to portray an action.

However, considering the role played by science in the history and development of modern civilization, it is difficult if not impossible to doubt the enormous importance of rationalism in science. Rationalism has been the central aspect in knowledge acquisition especially in modern science and also in the development of human epistemic awareness of the surrounding world. This has always been the traditional belief of the rationalist philosophers of modern philosophy like Rene Descartes and many others.

In the work edited by John Cottingham entitled, *The Cambridge Companion to Descartes*. In this work, the role of rationalism in knowledge acquisition has been properly emphasized and articulated about the enormous contributions of Rene Descartes; but lacks the role of relativity and quantum mechanics in modern science. He is seen by many as the most widely studied of all the great rationalist philosophers. His approach on the essence of philosophy and science tilted more attention on the role of reason in our every day understanding of properties and nature of the universe. Cottingham notes that Descartes' starting point in the quest for truth, his *Cogito ergo sum* ("I think therefore I exist") remains one of the most celebrated rationalist dictum of all time(1). This dictum is premised on the fact that he did not require any other material substance for his existence, but on the belief of the self. Thus, he contends that:

> From this I knew I was a substance whose essence or nature is simply to think, and which does not require any place, or depends on any material thing, in order to exist. Accordingly this 'I'-that is, the soul by which I am what I am- is entirely distinct from the body, and indeed is easier to know than the body, and would not fail to be whatever it is, even if the body did not exist (*The Cambridge companion to Descartes* 143)

Descartes, in view of the above, paved way for the growth of rationalism in science and philosophy within the modern era. His ideas proved the power of the faculty of reason than the material nature of the body, thus, by implication one could agree that he was not a materialist. He was motivated to believe in the self through his ratiocinative ability; this is because for him the quests for authentic scientific result using experience have always resulted in absurdity. According to Cottingham (2), Descartes stated thus: *Omnia semel in vita evertenda atqua a primis foundamentis denuo inchoandum* ("once in a lifetime we must demolish everything completely and start again right from the foundation"). This form of revolution is as a result of the inefficient result Descartes got from experiential knowledge and thus, paved way for what is known today as *Cartesian Rationalism*. For Descartes the driving force of scientific growth is laid on the foundation of rationalism. Critics would say, can rationalism alone give us an holistic understanding of the universe? Can certainty in science be found on rationalism alone? We agree in this research that, it is not the case that rationalism alone should be the final domain of scientific search, in other words, rationalism alone is not quite satisfactory in knowledge acquisition in science; as such the experiential aspect of verification through experiment and testing of fact will also be essential for scientific development, but rationalism is surely the leading role in the scientific quest for truth in modern science.

This is why Newton-Smith states that, the image that every scientific community likes to portray of itself always, and indeed the image that we must accept of that community, is that of rationality par excellence. He contends that the entire scientific community sees itself as a true paradigm of institutionalized rationality (*The Rationality of Science* 1). Smith's work provides a technical platform to view the scientific community as being propelled by reason. In other words, science is a product of rational creativity. He also agrees that within the history and development of science, scientists establish their discoveries first through theories, thus he states that the history of science is a tale of multifarious shifting of allegiance from theory to theory. This reflected mostly under Newtonian science, and later on the relativistic theory of Albert Einstein (3). This approach to science from Smith's perspective directs us to the rational explanation of science and agrees with the thesis of this research.

Roberto Torretti in his work titled *The Philosophy of Physics* also gave us another insightful explanation of rationalism in science. He gave us an illustration that reason performs a function by setting a goal in guiding our understanding of the universe in the process of knowledge acquisition and also regulates experience (138). He explains that the faculty of reason controls and directs our sense organs, consequently our understanding of the world of experience as well as the scientific world is driven by reason. Reason for him can be traced back to the ancient stoics, who called it the *hegemonikon* (the guiding principle). Torretti further argues in support of the stoics' notion of reason as the guiding principle for all scientific investigation, by stating that man's greatest power is the power to think rationally. Thus, rationalism as a guide implies that it gives direction and meanings to whatever the sense organs capture in the world of the phenomena which must agree to all that exists in the faculty of reason. Critics will inquire whether reason alone can guide us to gain indubitable knowledge of science. This will give Torretti a serious problem because of his total submission to the rationalist role in science; thus our research is an attempt to establish a lasting relationship between these two approaches of science and to adopt a new understanding that since the power of reason is enormous, as such should be regarded as the front wheel and should come first in any scientific investigation as pictured in modern science.

Nathan Spielberg and Bryon Anderson on *Seven Ideas that Shook the Universe* rendered an explicative approach on rationalism in science. They gave an elusive explanation of Galileo, Newton and Einstein's' contributions and achievement in science; which were all structured from reason. The book gave us seven of the most important and revolutionary ideas in physics. He contends that the world's greatest and successful revolutionaries did not just come up with ideas overnight; their work has brought lot of changes in science. Their successes have profoundly influenced modes of thought, actions and reinforced belief in reason and rationality, as operative tools of understanding the universe (2). The seven ideas according to Nathan Spielberg and Bryon Anderson include as follows:

1. The idea that earth is not the centre of the universe by Copernican astronomy.
2. The idea that the universe is a mechanism that operates according to Well-Established Rules by Newtonian Physics.
3. The idea that Energy Drives the Mechanism- the Energy Concept

4. The Mechanism Runs in a Specific Direction-Entropy and Probability
5. The Facts are Relative, but the law is Absolute-Electromagnetism and Relativity
6. You Can't Predict or Know Everything- Quantum Theory and the Limits of Causality
7. Fundamentally, Things Never Change-Conservation Principles and Symmetries (*Seven Ideas That shook the Universe*, 5-7)

Nathan and Bryon further argue that, the above listed ideas were all powered greatly through the tools of rationalism. These were defining moments in the history and development of science that was influenced by the spirit of rationalism. Reason according to the author(s) gave Copernicus the divine power to think outside the doctrinal motives of the church that the earth is but a minor planet among many planets orbiting the sun, thus the sun and not the earth is the centre of the universe. This rational power revived not only astronomy but the entire scientific world and even the church calendar, giving us the impression that through creative reasoning the impossible can be achieved.

The second revolution came from the successes achieved by Isaac Newton, one of the most celebrated scientists of all time. He posited that all objects in the universe are subjected to laws of Physics; laws of Motion and Law of Gravity. This idea recorded enormous successes in science and reduced science into a precise mathematical formula, and further proved that through reason more can be achieved in science. These laws were comprehensive and comprehensible and vividly explained all the activities of the macro world, up till today these laws are still in operation and his ideas on mathematics are still crucial in most scientific experiment.

One of the theories that support rationality in science is Quantum mechanics, the theory brought into science a defining moment were reason became a tool for scientific justification. *Oxford Dictionary of Chemistry* defines Quantum mechanics as that branch of physics which deals with the behaviour of matter at the level of the atom, the nucleus, and the all elementary particles. At this of level of reality, energy, mass, momentum and other quantities do not vary continuously, as they do in the large-scale world, but come in discrete unites called quanta (986).

Quantum mechanics as a mathematical theory, as a replacement for classical mechanics in order to explain satisfactorily the behaviour of atoms, molecules, and elementary particles in terms of observable quantities such as the intensities and frequencies of spectral lines. This explanation gave us a picture of what this research is yet to investigate on concerning the place of reason in modern science.

Quantum mechanics changes or revolutionized the understanding of nature at the time it was discovered by Max Planck in 1900 (Brennan, 93), and up till this present time its successes are enormous and have also reinforced rationalism in science at its fullest. Most discipline have today adjusted their syllabus in-line with quantum mechanics and its predictions, this is why we, have, quantum chemistry, quantum electronics, quantum optics, quantum logic, quantum information science and many others.

On the account that, Quantum mechanics gave a clear and rational explanation of what Newtonian empirical science could not. Egbai, Uti in his work titled, *The Emergence of Subjectivism in Physics and the Possible Implications* published in *Sophia: An African Journal of Philosophy and Public Affairs*, noted that, Neils Bohr and Werner Heisenberg the chief contributors and foundationalist of quantum mechanics on the Copenhagen meeting avers that, quantum mechanics is "the end-of-road-thesis in science", the final theory, the never-to-be surpassed revolution in physics (14). This implies to Bohr and Heisenberg that no other theory will supersede the rational powers and predictive energy of quantum mechanics.

Egbai, Uti further explain the rational thought of Einstein relativity, when he avers that, it is first of all important to remember that Einstein never regarded any of his own theory as a final breakthrough. His own photon theory and the need to use it together with the wave theory of light, which really established what was later called wave-particle duality, he did not regard it as a stop gap, although it brought him almost to the threshold of the theory of matter wave; His special theory of relativity he also regarded, rightly as unsatisfactory for several reasons, empirically because it merely replaced absolute space by the absolute set of inertial systems (17). Egbai here attempts to capture the rational image of Einstein relativity and how Einstein thought about the growth of science, giving space for full application of reason in the development of science.

On the rationality of quantum mechanics, John Archibald Wheeler on *The spooky quantum,* posits that quantum mechanics explains that every physical object emits electromagnetic radiation with a smooth wavelength spectrum that depends on the temperature of the body in the case of object from every day experience, such as rock or a human body, that radiation is in the infrared region of the spectrum where the wavelengths are larger than those of visible light and not detectable by the human eye. So in the absence of any reflected light, it appears to the naked eye that the body is black (9). This also gives us a good outlook on how quantum mechanics operate in everyday reality concerning black body radiation. As such, within quantum reality we are limited to what we think and not what we see, because subatomic particles are unseen element of nature.

Nielson Rud on *Neils Bohr Collected Works* also gave us background knowledge of quantum interpretation when he contends that in quantum theory objective reality has disappeared and the reality which brings new theory has evaporated as one has no fact to gather which will help one to arrive at a new theory. He argues that the empirical objective world of fact gathering exists no more at the quantum level and the idea of causality also ceases to exist since quantum particles do not obey the laws of nature. Neilson did not consider that the methods of empiricism in science are still applicable in today's science as such his idea does not capture science holistically. But his idea shows the power of rationality in science.

According to Karl popper in his book *Quantum Theory and the Schism in Physics,* he gave us another clear picture of quantum mechanics and its relationship with relativity as two rational discoveries. Thus, Popper admits that:

> Bohr too was, of course, a passionate admirer of special relativity theory. He would have wanted to avoid rejecting it, like almost everybody in those days. Had it been shown that such rejection would be necessary if we wish to uphold quantum mechanics, it may well have meant, even for Bohr, the rejection of quantum mechanics. For special relativity more or less set the standard to which quantum mechanics had to conform. (30).

The view from Popper gave us an illustrative background on the relationship relativity and quantum theory had, in attempt to proffer solutions in science through the theory of everything; an attempt that has richly become the base of modern rationalist moment in science called string theory.   These are all scientific theories that are projected by rationalism

However, a related work is Chris O. Akpan's research entitled: *"Quantum Mechanics and the Question of Determinism in Science".* He exposes holistically the concept of determinism as a Newtonian approach to science which was later overthrown by the birth of quantum mechanics. He argued that: Quantum mechanics cannot totally be free from determinism, for it has subtly conceded to determinism in holding that the sub-atomic particles are the building blocks of the macro world, and that there is interconnection between entities in the universe, even at the non-manifest level. The probabilistic interpretation of quantum mechanics only belay our human limitation of the knowledge of nature, which I believe still have lots of tricks in our sleeves over mankind (72). The author attempt to argue on the bases that science should not only be seen as a rational discipline of quantum mechanics as such. For him the mechanistic law of the universe is still at play by scientists. We agree with this author, but our point states that rationality is more in modern science than empiricism.

John Stuart Mill in his book entitled, *System of Logic Ratiocinative And Inductive: Being A Connective View Of the Principles of Evidence and the Methods of Scientific Investigation,* he attempted to solve the problem of scientific method by introducing a method of studying science called the *Ratiocinative And Inductive.* Here he stated that "Every induction is reducible to deduction and all deductive sciences are inductive" (164). By implication science should be seen as a discipline that obeys experiential and non-experiential aspect of the universe. This idea is what influenced our deeper understanding of the nature of reality and led us to approach science within the modern era as a product of rationalism.

Another critical position in the search for the method of science is the one postulated by Paul Fereyabend. In his book, *Against Method,* he attempted to fashion a new model of scientific method, but ended up stating that, the whole notion about the method of science is an illusion. For him the method of science is full with enormous limitations and as such "anything goes". This is so because according to Fereyabend the idea that science should have a universal fixed rule is unrealistic and pernicious and the only 'rule' that survives is 'anything goes' (296). This work, rejects the aforementioned idea of 'anything goes' because science operates on fixed principles that needs a dynamic method to understand its operation in the universe.

Immanuel Kant is another philosopher of science whose effort in the search for scientific method should not be forgotten in a hurry, in his *Critique of Pure Reason,* he postulates a new model for all scientific research called the *synthetic a priori* knowledge. Here Kant asserts that scientific knowledge is not dependent on what is presented in experience alone, but also through the application of logical reasoning. This was also an attempt to unite empiricism and rationalism. But the question one may ask in Kant philosophy is that, which of these methods should be seen as the front wheel for science. It is imperative to note that despite Kant attempt to unite these two methods, he failed to articulate what could possibly lead science as the foundation of research. Thus, moments in science today have proven the rationalist content of science which has exposed the weakness of Kants' position, which tilted more to empiricism.

In expounding the notion of reason in science, Austin Fagothey in his work *Right and Reason: Ethics in Theory and Practice,* gave an explicit account of the role of reason in the justification of human knowledge and actions. He started by addressing four questions in the search for ratiocinative knowledge. He asked the following questions:

1. When is human reason right?

2. How does right reason act as an ethical norms?

3. How is right reason applied to human nature?

4. How practical is right reason as a norm? (99).

In response to the above questions, Fagothey argues that human nature permits reason to function within the area of raw data gotten through experience. He argues that reason do not function in a vacuum, but need some data formed through experience and human conduct (100). He further explains the two roles of reason in the formation of human norms; the first which is reason as a rational exercise and the other; reason as irrational exercise. According to Fagothey, reason that is used to plot crime is irrational, while reason that is used to correct crime is rationally based. Thus, the promotion of human capacity to reason is a distinctive part of human being. He further admits that, because of man's critical approach prominence is given to reason; the inability to construct a complete rational system is not because of the incompetence of reason or a matter of unintelligibility, but due to the slow application of man's intellectual force (101). The strength of Fagothey's position proves the inevitability of rationalism in every human action, the driving force of human norms is practically connected to reason. This research supports such a great agreement of reason and human action, but we argue that through reason, human beings are fully incorporated to the nature of the universe of science.

William F. Lawhead in his book entitled *The Voyage of Discovery: A Historical Introduction to Philosophy,* introduced us to the great works of many scholars whose effort are monumental in the quest for knowledge. One of these works is the contribution developed by Gottfried Leibniz whose effort on the concept of monads has influenced the development of rationalism in modern science. Leibniz agrees that the fundamental unit of every reality is what can be known as monads. The concept derived from a Greek word 'Monas', meaning unity or 'that which is one'. According to Leibniz the entire universe is made up of infinity of simply, nonmaterial, invisible substance called monads (*The Voyage of Discovery* 265), and these monads is what make up the formation of every reality. Lawhead further describes the monads as possessing the quality of self-enclosed and self-sufficient unit; meaning nothing enters and nothing goes out of a monad, this is what Leibniz described as the windowless monads (265). This great concept added much theoretical framework in chapter four of this work. In science, the development of subatomic particles is derived from the idea exposed by Leibniz and have also produced a vital force for the development of quantum theory in physics.

Kyrian A. Ojong in his book entitled *A Philosophy of Science for Africa,* gave an exposition of some philosophers of science whose efforts was to produce a method for scientific research. He critically examined scholars like Karl Popper, Imre Lakatos, Thomas Kuhn, and Paul Feyerabend. These scholars in their efforts attempted to produce a solution for the quest for scientific methodology and to solve the problem of demarcation between science and non-science. The author also examines the various criticisms of these scholars which have proven that the search for scientific methods is a continuum. In the introductory section, the author admitted that scientific research proceeds not just by one method but diverse methodologies (3), to this effect the continent of Africa can still produce a scientific model that could lead to the growth and progress of science in Africa. This is because, according to the author, the African continent record a very slow pace of development in the sphere of science and advancement in technology, thus his aim is to establish on a peculiar philosophy of science that suit the African approach to scientific epistemology (4). He argued that the African scientific epistemological search goes beyond the empirical reality, thus, African believes in two realms of reality, the empirical and the non-empirical. As such a monolithic understanding of reality may bring scepticism to science. The strength of this work lies on the exposition of some philosophers of science whose ideas contributes to the theoretical foundation of rationalist philosophers in modern science as discussed in chapter four of this work.

Chrysanthus Ogbozo in his book entitled *Philosophy of Science: Historical and Thematic Introductions* gave a full exposition of the historical introduction of Philosophy of science. The book composed of eight chapters aim at bringing an historical survey of the various philosophical epochs and how its effect human civilization. In chapter one, the author examined four philosophical era; the ancient era, medieval era, renaissance era and contemporary era; which according to the author had so many impacts in the history of scientific development. In chapter two, Ogbozo critically exposed the various problems, meaning, and nature of philosophy of science. The chapter further introduces basic concepts in scientific studies that have relevance in philosophy of science.

In chapter three, a greater consideration of major scientific methods was discussed, methods like Pythagorean/mathematical method, inductive-deductive method, experimental method, A Priori deductive methods and analytic-synthetic method where all examined. In chapter four, the author contends that reality do not manifest in one way, in other words there are so many approaches in studying the universe of science. In chapter five, the author attempts to explicate on the notion of scientific truth, the meaning of paradigm, and the theories of science. He defines a theory as a system of ideas which explains something especially on that is based on observation or on general principles (168). This definition of a theory has an empiricist colouration alone in it. The weakness of this definition is that it lacks the background of the unobservable theories of science like quantum theory. In chapter six, the author subjected certain scientific ideas to philosophical scrutiny, giving us the motivational impacts of philosophy as the mother of all discipline, an aspect that is considered as a vital force in this research. In chapter seven and eight, the author shows great power on the inductive principle of science; thus, the quest for scientific knowledge according to these two chapters gave the background of a science that is derived through particular to general; in other words, from a particular observation to the general law as also posited by Chalmers in his book *What Is This Thing Called Science*. Here Chalmers contends that scientific laws start from observations of particular instances through which they become a general law (Chalmers, 8). However, the strength of *Philosophy of Science: Historical and Thematic Introductions* lies on the power of its ability to excavate the historical nature of science from its philosophical perspective. The book gave this research the literally understanding on most of the vital objective of philosophy of science. Though the book did not cover contemporary issues in philosophy of science, but its content added a vast literal understanding to this research.

In conclusion, from the above works reviewed, no author has rightly justified how quantum mechanics and theory of relativity contributes in the formation of rationalism in modern science. This work therefore, is justified by its insightful exploration of moments of formation of relativity and quantum mechanics in the growth and development of rationalism in modern science.

CHAPTER THREE

RATIONALISM AND EMPIRICISM IN MODERN SCIENCE

This chapter focuses on background to modern science, empiricism in modern science, rationalism in modern science, and finally the rationalist-empiricist debate in modern science.

BACKGROUND TO MODERN SCIENCE

The motive behind this background to modern science is to give an historical analysis of what gave birth to modern thinking and the major players who made it possible for science to become a systematic discipline.

Bertrand Russell in the *History of Western philosophy* identifies four great men as among the Pre-eminent scholars in the creation of modern science. They are Copernicus, Kepler, Galileo and Newton (512). Though Copernicus' contribution was within the sixteen century, but his heliocentric theory put forward the challenge for the rise of science and to explain the anomalies of the sixteen century (Akpan 113). Nicholas Copernicus is seen by many as one that started the investigation against the orthodoxy of the church by bringing in a revolution that gave birth to the heliocentric cosmology; a hypothesis that established the sun as the centre of the universe, and that the earth rotates around the sun (*The History of Western Philosophy 513*). One of the difficulties that this theory raised was that, it contradicted the bible. Others were seen from the perspective of measurement of the earth within its rotational realm with the sun and the distance with the stars. It was also difficult to scientifically explain the position of falling objects during the rotation of earth around the sun. This explanation was possible during the time of Galileo's discovery of the laws of inertia. But in the time of Copernicus, no answer was forth coming (*The History of Western Philosophy 515*).

We must give merit to the men who founded modern science because these men were immensely patient and humble during observation, and had great boldness in framing hypothesis despite the challenges that came their way. Copernicus possessed a greater humility & simplicity as an Astronomer. In all, he proceeded with great zeal by observing the motions of the heavenly bodies patiently and with great humility.   Up till today, Copernicus is celebrated by many because of his bold steps in finding fault on what has been believed since ancient time as true, as this was possible through collection of facts. He was indeed a great participator in the rise of science.

In the first quarter of seventeenth century, Francis Beacon introduced the quantitative inductive method which reduced scientific arguments to a set of rules and regulations. In the second quarter Rene

Descartes responded with his deductive-mathematical mode of reasoning. This deductive mathematical method came in as a response to Beacon's quantitative induction to examine science and philosophy as a product of deductive regulations (Akpan 113).

Johannes Kepler who was born on December 27, 1571 and died on November 15, 1630 (Ephraim 235) is today respected for his contributions to the development of modern science. He was a notable man in science because he took observation as a valuable tool which influenced the growth of Astronomy in science. He was the first astronomer after Copernicus to adopt the heliocentric theory. One of his greatest achievements was the discovery of the three laws of planetary motion in 1909. These laws are stated as follows:

1. The planets describe elliptic orbits, of which the sun occupies one focus

2. The line joining a planet to the sun sweeps out equal areas in equal times

3. The square of the period of revolution of a planet is proportional to the cube of its average distance from the sun (*The History of Western Philosophy 515*).

These laws gave birt   h to a modern thought system for a modern man to understand the orbiting nature of the earth around the sun.

Galileo (1564-1642), is seen by many as the greatest of the founders of modern science outside Newton. He died the year Newton was born. He was greatly into dynamics (A force that produces motion). This led him to first discover the importance of acceleration in dynamics, as such; any change in motion of bodies is as a result of force acting on it. This principle was later interpreted by Newton in his laws of motion.

Isaac Newton is another great scholar that gave meaning to science and created a science that is justified by reason. Modern Science was totally influenced by the works of Isaac Newton. He was an intelligent scholar and indeed a chief advocator of modern systematic thinking. The era of Isaac Newton was remarkable and fashioned in objectivity, causality and determinism to all scientific knowledge. These three assumptions in modern science help to systematize science and compel scientists to abide by a remarkable standard all over the entire community of science. Isaac Newton reduced science into precise mathematical laws capable of pushing in objectivity and shaping the quest for scientific knowledge with the aid of reason. His laws explain everything about the world of matter (Macro world) and everything ever seen using sense perception was all explained by the laws of Newton. These laws are what Newton called the three Laws of motion and the universal law of gravity.

Empiricism in Modern Science.

Modern science is an era of scientific investigation that was influenced by objectivity and inter-subjectivity through observation and experiment. Through this background, the method of induction became the starting point of scientific movement as supported by the empiricist notion of science. This inductive method is the inference of a general law from particular instances gotten through observation or inferring universal statements from singular ones no matter how numerous the observation instances may be (Ojong 9). Thus, Andrew Uduigwomen in his thought provoking book gave us a vivid illustration of the nature and scope of empiricists' notion of science within modern science. Science for modern scientists according to Uduigwomen is a kind of knowledge arranged in an organized or orderly manner, especially knowledge derived from experience, observation and experimentation (*A Textbook of History and Philosophy of Science* 20). It is through observation that scientific method of induction is made possible.

However, up to date, many philosophers and physicists still engage themselves on the puzzle of what could stand as the method of scientific research. This great puzzle is in attempt to unravel how viable science can solve the ultimate problems of man and his physical universe. To this effect the tradition of empiricist's minded scholars pursue natural understanding base on the aspect of reality that could be ascertained only from pure empirical analysis of the universe or a science that could be justified through experiment, paving way for knowledge gathered through sense data collections as the major means of acquiring authentic indubitable means of knowledge. Science for the empiricist is seen as an epistemic awareness that seek to study the nature, scope and origin of the universe through the aid of human experience. The philosophical question one could ask is how viable is our empirical knowledge about science or reality in general? Can empiricism alone in science serve as a true method for scientific research?

Physicists and philosophers of physics all over the world in various ways are greatly involved in this investigation, and as such attempt to articulate their contentions on how reliable human experience alone could help to achieve a long standing solution that could help science unravels the mysteries behind the nature of the universe. Within these meticulous investigations, philosophers of science do come across difficult challenges due to human limitations of capturing reality through the aids of sense data collection, such challenges includes the wrong confirmation of a pool of water in a good road which ends up to be a mirage, the wrong perception of a curved stick, when a straight stick is inserted into a bucket of water, and most time answering to voices when indeed no one calls due to hallucination, others of the same category are forms of delusion, deception, illusion and phantasmagoria (Ozumba 83) as the endemic problems surrounding the issue of empiricism in science . These and many more are some of the ways our sense organs do deceive in our attempt to perceive and understand reality and thus when what we sense do not agree to reality, it is therefore paving way for spontaneous scepticism in science.

Despite the enormous success achieved by empiricism in science, through the contribution of Isaac Newton's laws of motion, law of gravity that help to mechanise scientific ideas and other empirical minded scientists, empiricism alone has enormous limitations as previously explained. Thus, the problem of what could constitute the leading method of scientific research has become an unsolved issue in science; also giving much aid to empiricism has attracted much criticism from the rationalists minded scientists. These have all brought in scepticism in science to create a new dimension and a method for scientific research that could skip a greater part of scepticism in science. Thus, reality cannot be seen as the handmaid of empirical analysis alone because of its empirical limitations in understanding and coping with the challenges and unravelling the mysteries behind the universe.

However, empiricism in science up till today has suffered a lot of criticism both from scientists and philosophers of science because of its enormous limitations. The questions that may puzzle critical minded scholars like Paul Feyerabend are; can only the observable be studied scientifically? Can observation solve all humanitarian problems in science? How certain can observation inform us about the true nature of the universe? Can scientist derive meaning from science through what they observe alone as the foundation? In other words, what could be the foundation scientific research? These are few of the thought provoking questions that a philosopher of science may be tempted to inquire about empiricism in modern science.

In reaction to these topical questions, Paul Feyerabend in his thought provoking book entitled *Science without Experience,* gave us a critical exposition of the limitations of empiricism in science. He started by asking pertinent questions about empirical hypothesis, whether the empirical hypothesis is correct, that is, if experience can be regarded as the true source and foundation (testing ground) of knowledge? For him a science without the aids of experience is a possibility. Thus, it is possible for us to examine a scientific investigation without sensory element (*Journal of Philosophy* 791). He demonstrates that there are three cardinal points or processes in science in which experience is irrelevant. These points includes: testing; assimilation of the results of testing; and understanding of the theories (791).

Science from Feyerabend's perspective can be investigated without the help of experience, because for him theories are just strings of signs without relation to the external world, unless we now design mechanism of connecting them to experience, experience will not be necessary. Testing of fact may not need experience before one can test a particular theory; it is the work of human ratiocination to communicate in the process of testing and to even understand the contents of a particular theory after testing. According to Feyerabend, it is easily seen that experience is needed at none of the three points listed above (791). In his criticism, he further noted that experience arises together with theoretical assumptions, not before them, and that our experience about the natural universe without theories is just as uncomprehended as is a theory without experience. This is what he termed as observational-theoretical dichotomy. The information is that, for Feyerabend observation should not come before theoretical knowledge of a given reality; as such theories which are rationally based should serve as the foundation of our notion of science. Because knowledge for him can enter our brain without being contacted through sense organs. The question here is how? He submitted that, the mind has the power to create knowledge through deduction of ratiocinative reflections that do not connect with experience which may be gotten through intuition.

Empiricism also has other enormous limitations in relation to modern science. These limitations are revealed in the moments of the formation of relativity by Albert Einstein, formation of quantum mechanics through Max Planck discovery of Quanta in his Blackbody experiment, and moment of string theory. One can insightfully capture that empiricism in science has a lot of deficiency in-view of the above listed moments of science. In relativity one needs a high level of rational reflections to understand the nature of light, energy and mass differences which gave birth to the equation $E=mc^2$. The equation that proves the equivalence of mass and energy are not distinct but are two forms of same thing. In this equation E symbolises Energy; M is for Mass; C is for the Speed of Light and 2 for "Squared". Thus, Einstein did not experience the idea of energy-mass equivalence before proposing the equation; it was an aspect of what he called *thought experiment* and not through physical experimentation (Marion 462; Brennan 84), thus this theory revolutionized both science and philosophy (Bryan Magee 220).

Another limitation of empiricism in modern science is on quantum mechanics. Quantum mechanics which studies the unobservable parts of reality do not obey empirical laws of Isaac Newton. The nature of quantum mechanics cannot be proven mechanistically nor can the ideas of dark matter and dark energy be accounted for using the laws of Isaac Newton or through any other experiential media. This is why empiricism has become a misfit in the micro world of subatomic particles.

Richard Brennan in his work *Heisenberg Probably Slept Here* agreed that classical physics could not solve the problems of micro world, when he rightly contends that:

> Strange phenomena occur in the world of the very small. One of the most difficult to understand is wave-particle duality. Classical physics makes a clear distinction between a wave and a particle. But in the realm of the very small, these distinctions become blurred. Numerous experiments have shown that in the strange world of atom, a physical entity somehow manages to possess a dual characteristic, sometimes appearing as a particle and sometimes behaving like a wave. A pinpoint particle and a spread-out wave seem to be two quite distinct concepts, but in the subatomic world the two seem to merge (104).

It is pertinent to note that Planck and Einstein established the wave-particle duality of light, but they did not realize that this concept could be extended to all subatomic particles. Quantum theory through rationalist approach explains the activities of the subatomic particle, and gives us an enormous explanation of the behaviour of wave and particle in the concept of wave-particle duality, meaning that light (photon) can sometimes behave as a wave and can also behave as a particle, depending on what the observer chooses to observe (Brennan 104). But classical mechanics could only account that it can be wave or particle not knowing that it can both merge. Thus, Brennan further noted that through rationalism in quantum theory, the strange world of atoms and physical entity somehow manage to possess dual characteristics, sometimes appearing as a particle and sometimes behaving like a wave (Brennan 104). Because of the limitation of empiricism in classical mechanics, Shashi Chawla agreed that the classical theory was unable to explain the following experimental observations regarding small objects; which include the following:

i. Discrete spectra emitted by excited atoms

ii. Photoelectric effect

iii. Variation of heat capacity of mono-atomic solids with temperature and, Spectral distribution of energy in blackbody radiation etc. (A *Text Book of Engineering Chemistry* I).

This implies that empiricists' method alone became less reliable in science. Scientists needed to solve these problems, and they had to pave the way for quantum mechanics where rationalism reigns and where the problems of subatomic particles could be explained. This is why this research maintains that rationalism is the foundation or the stronghold of modern science.

Rationalism in Modern Science.

This section centres on rationalism in Modern science. The idea of rationalism came into philosophical discourse to portray a kind of belief system that gives credence to innate ideas as the root of all knowledge acquisition. It can also be seen as a philosophical school that posits that authentic knowledge comes to us only through reason. And it is only through deductive ratiocination that one can get a good grasp of the entire nature and scope of the universe. The rationalists in their own dimension seek to understand nature or the entire universe of science through the aids of reason and deduction.

Godfrey Ozumba contends that any knowledge that is based on sound reasoning, logical deductions, mathematical procedures and other extra-sensory means are termed rational knowledge (*A Consise Introduction to Epistemology* 59). In science, the application of mathematical rules is at the heart of rationalism in scientific discovery; scientific theories are first of all confirmed using mathematic symbols. These symbols represent the application of pure intuitive cognition of abstract deductions gotten through reason. Thus, the rationalist model of science is a direct opposite of empiricism, and thus believes that since empiricism fails in giving us an indubitable knowledge, rationalism therefore is the foundation and also the only solution to scientific method that can explain the new dimension to scientific knowledge. In various approaches, the rationalist's moment has been a major motivation that has forced in the ideal substance of rationalism in science. Such moments include the moment of formation of theory of relativity, quantum mechanics and even string theory. Today, enormous successes have been achieved from these three theories in science and have being the foundation of modern development in science and technology. The production of Magnetic Resonance Imaging (MRI) used for brain scanning in modern hospitals, the fast internet connections, the invention of modern wireless telephone devices for easy communication, the invention of radars for airplane tracking through Global Positioning System (GPS) network, our remote devices at home for digital televisions, introduction of lasers for modern electronic printing, phones and cinemas and even explosive devices used for international security. These are made possible in science through the predictive power of theory of relativity and quantum mechanics, moments that have lifted science through data gotten from consciousness or activity of the mind over experience.

In rationalism, reality is analysed through innate justification, just as Paul Fereyabend predicted in his work entitled *Science without Experience* and noted that a science without experience is a possibility. This idea gave science a vital force to solve most of the problems our senses could not solve. Karl Popper also agreed with Fereyabend's notion when he admitted that "physical science cannot discover the hidden essence of things" (*Conjectures and Refutation*, 425). This assertion of Popper is based on his belief that the essence of a thing depends on a deep understanding of the unseen reality as captured through ratiocination. This is what gave birth to a shift from Newtonian mechanics to Quantum mechanics which gave a necessary explanation of the unseen realities in science, because reality needed to be examined and interpreted outside experience to solve the fundamental problems of the unseen universe.

However, rationalism also has been a major force in science, but cannot be the only method for science because reality consists of empirical and rational components. But because of the nature of science, rationalism plays a foundational role in modern science than experience. In other words, empiricism alone cannot give science a holistic view of the total description of science, since rationalism is what maintains the leading role or the foundation in any scientific research.

Furthermore, modern science celebrated much of rationalist contents because the idea of observation, experimentation and verifiability criteria of the logical positivists needed to be justified by intelligibility before acceptance. This is why the tenets of rationalism in science became necessary in theory formations. They are as follows:

1. Reason is the foundation of scientific knowledge.
2. The power of science is only derivable through ratiocination as the building block of science.
3. The method of science is the method of deduction.
4. All knowledge comes to man through innate pulses and are justified through rationality before experience.

It is through the utilization of these tenets of rationalism and ratiocination that scientists gain the understanding of relativity science and quantum mechanics. This is why this work noted earlier that quantum mechanics is built on the foundation of rationalism. Thus, John Barrow corroborates this view when he states that, we cannot achieve the very high energies needed to unlock the secrets of elementary particle world by direct experiment (Theory of Everything 39). No scientist has ever heard the sound of a photon, or felt the weight of mesons; neither has any ever felt an electron or photon. This part of physics can only be studied by uncertainty and indeterminacy principle and justified by reason. The Uncertainty and indeterminacy principle stated by Neils Bohr and Werner Heisenberg posits that the more precise you determine the position of subatomic particles the less precise the momentum can be gotten. This theory shattered the foundation of classical mechanics by eradicating absolutism in physics because at the level of quantum mechanics the deterministic laws of Newton ceases to exist (Egbai 105).

The String theory in science is another important moment that has given us a new paradigm of rationalism in modern science. The theory has been proven mathematically through rational deductions of mathematical rules. According to Andrew Jones, the theory states that reality can be reduced into some tiny loops of vibrating energy called strings. String theory currently is the only theory vying for the total explanation of all known entities, namely, the world of matter, theory of relativity and quantum mechanics. This theory predicts that all forces of the universe could be unified into a single manipulation of tiny loops of vibrating energy called strings (*String Theory for Dummies* 12). The long quest for this theory has been a demanding development in the search for the theory of everything by modern physicists which started with the works of Albert Einstein to unify his theory of relativity with quantum mechanics called quantum gravity (Hilgevoord 42), and made it possible to view reality as one dimensional string. This theory predicts the existence of dark energy, dark matter and subsumes everything in the universe into tiny vibrating unseen strings. Thus, its reality remains uncertain but has been proven to be in existence only through the aid of mathematics. This moment of science has validated Feyerabend's earlier notion that science without the aids of experience or observation is a possibility (*Science without Experience* 794).

Therefore, it is true that all these moments of quantum mechanics and string theory are moments in science that neglect observation and experience as the foundation of scientific knowledge. Thus, the idea that rationalism has always been the foundation of scientific growth is what we call the ratio-empirico-centric approach which supports the rationalist content of science as the front wheel during any scientific investigation about the nature of the universe.

The Rationalist-Empiricist Debate in Science

The debate between rationalism and empiricism in science has always raised one of the thought provoking issues in philosophy of science. The empiricists till today believe that science is an empirical dependent discipline, and anything called science should pass through strict experimentation from fact gotten through raw data of experience. Conversely, the rationalists state the opposite, and posit that, scientific investigations are usually based on deduction derived through the process of ratiocination. This heated debate originates in attempt to rightfully justify the foundation of science.

In physics specifically and in the sciences generally, Isaac Newton drew out a "clock-work" universe that helps to solve most of the challenging issues in science by establishing his laws that brought in certainty to science. Newtonian science helped to define science mechanistically using the collection of reason and experience. His laws of motion and gravitation all proved enormous success in physics and mathematics (calculus), also gave rise to the philosophical implication that, absolutism is another way nature can manifest through fixed laws based on observation and our experience of the universe. But the theory of relativity, quantum mechanics and string theory have come to create another unique way of understanding the universe, where reason through the tools of consciousness can create reality outside experience called qualia (Deepak & Menas 287). This is a remarkable moment that shows that science is not empirically dependent as posited by most empiricist scholars.

In the work of Asouzu, he argues that, reality is better understood when viewed in a holistic dimension and all manifestation of the universe from both rational and empirical strains must be conjugated in some cases, but the application of reason as the foundation of any scientific investigation must be considered. In other words, the visible and invisible, subjective and objective, physical and non physical, experiential and non-experiential must be studied as part of science that should make up a universe of science. This is so according to Asouzu because; the conception of a holistic reality calls for a complementation of all existing factors. Thus, Asouzu describes reality as the complementation of all missing link and missing links here are the various manifestations of reality in units, parts and in different compartments (*Method and Principles* 285-286). The manifestations of reality according to Asouzu exist in various units such as, the essential and accidental, the necessary and contingent, the universal and the particular, the absolute and relative, the conservative and the progressive, the constructive and the deconstructive, the real and the ideal, the transcendent and the world-immanent, the empirical and the non-empirical, all these can easily be grappled within the same frame work of ratiocination (*Method and Principles* 285).

Thus, taking a close look on the various manifestations of reality, it is rather rationalism that takes the foundation of the entire gamut of science through the aid of innate ideas. Let us examine some issues that need reason as the foundation of knowledge. If two people are observing the colour of water in the sea, they may choose to disagree over whether the colour is green or blue, but will not disagree on what "colour" means, this is where reason plays a vital role. They may also decide to taste a bitter cola and argue over whether it is bitter or not but they won't disagree that taste is involved. They may also observe the dominant colour of the Nigerian national flag, and may argue whether it is green or white and not whether colour is involve. In another case, they may also choose to observe the behaviour of light, one may say it appears like a particle and the other may argue that light is actually a wave. Here both views are rationally correct because from the background of wave-particle duality of light as explained by rationality in quantum mechanics, water, sound and any other physical entities are made up of wave and particle called wave-particle duality (Brennan 104). Critics may ask, can blind people know colour? Or is experience not crucial in knowing colour? Others may ask, can a person without taste buds know taste? The answer is no, because there is a missing link between the blind and the physical universe. The question is that modern science does not consider true knowledge as empirical dependent, because science at the modern era proves that the truth about nature has tilted away from empiricism to rationalism.

Thus, in the purview of the above instances, if these two observers argue against "taste", "colour" "wave and light" they will surely be trapped in a paradox. Therefore, in understanding reality holistically, the role of reason produces much understanding in scientific investigation because the remarkable moments in modern science were all possible by rationalism through the aid of thought experiment. This work further states that there is no one way of perceiving reality or carrying out scientific investigations and avers that a recourse to reason or rationalism rather than empiricism should serve as the foundation of science. The empiricist definition of science as empirical dependent is not a suitable way to explain this understanding because without rationalism in scientific studies there will exist a large vacuum in understanding the way nature unfold. This is why Karl Popper admitted that the essential character of reality cannot be cognate through physical science, in other words, for Popper the physical science cannot discover the hidden essence of things (*Conjectures and Refutations* 104).

Buttressing the role of rationalism in science, Einstein agreed with Popper when he admitted that the whole of science is nothing more than a refinement of everyday thinking (Bryan Magee 220). From the foregoing debate, this work subscribes to the foundational role of rationalism and admits that the process of ratiocination yields more discoveries in science as shown in the theory of relativity and quantum mechanics.

CHAPTER FOUR

EXAMINATION OF SOME RATIONALISTS IN MODERN SCIENCE

In this chapter, we focus attention on the works of Karl Popper, Imre Lakatos, Gottfried Leibniz and Isaac Newton. This work examines these scholars in an attempt to build support and to articulate the tenets of rationalism as foundational in the formation of modern science whose contributions have also helped in strengthening scientific development within the modern era.

Karl Popper's Critical Rationalist Methodology of Science

Karl Popper was born in Austria-Hungary now Vienna on 28 July 1902. At the age of 16 Popper attended lectures in mathematics, physics, philosophy, history of music and psychology as a zealous student at the University of Vienna. He is regarded by many as one of the greatest philosophers of science in the 20th century and a denoted opponent of all forms of scepticism in modern science (Akpan 38). Popper had a strong quest to solve the problem in philosophy of science; the demarcation problem that presents how to distinguish science and non-science or pseudo-science was one of his greatest challenges in science. This gave birth to what he called the critical rationalist methodology of science. He has published great books as a credit to his scholarship, some of which are *The Logic of Scientific Discovery* (1934), *The Open Society and Its Enemies* (1945), *The Poverty of Historicism* (1957) and *Unended Quest* (1976).

The term "critical", according to Uduigwomen, means faultfinding or forming and giving a judgement or opinion on something, and rationalism here denotes the practice of treating reason as the ultimate authority in all subjects of study including science (228). This methodology of science, according to Popper, is an attempt to distinguish rational thinking from irrationality by examining the problems critically. Thus, Popper states:

> And yet, I am quite ready to admit that there is a method which might be described as the one method of all rational discussion, and therefore of the natural sciences as well as of philosophy. The method I have in mind is that of stating one's problem clearly and examining its various proposed solutions critically (*The Logic of Scientific Discovery 16*).

Popper was of the opinion that the growth of science depends on the ability of being rational and critical and that the foundation of science depends completely on the ability to apply reason in any scientific investigation and not through induction. Induction, in Poppers view, is not a criterion of science. This is why he set out to investigate the following:

i.   The solution to the problem of induction
ii. The problem of demarcating science from non-science, and
iii. The importance of maximizing criticism and maintaining a critical attitude as essential for rationality as vital for the growth of knowledge. (Ojong 41)

Thus, Popper had the intellectual courage in changing his mind under the influence of criticism that is rationally based. Popper's quest to solve the demarcation problem can be conceptualized under the following themes:

1. The Concept of Falsifiability Criterion
2. The Concept of Verisimilitude (Truth-Likeness) and
3. The Concept of Corroboration

The falsifiability or refutability criterion is one of the pillars of Poppers methodology of science. According to this view, the credibility of a theory to be called scientific depends only on if it makes assertions which may clash with observations and a system is in fact tested by attempts to produce such clashes; that is to say, by attempt to refute it. Thus, for Popper, testability is the same as refutability and this can be taken as a criterion of demarcation (*Conjectures and Refutation* 256). This is the only critical approach that can be used to demarcate science from non-science according to Popper. By testing it, in order to know if it can be critically discussed, through which its limitations can be deduced. During this process, Popper noted that some theories expose themselves to possible refutations more than others because of their degree of testability (*Conjectures and Refutation 257*). This testing ability is what makes science to grow and he called this entire process "conjecture and refutation" which means trial by error. Popper also used the term verisimilitude. Verisimilitude means, the truth content minus the falsity content.. Rather than devise a theory of realism founded on a doctrine of truth, Popper had devised a concept called verisimilitude by which the truth-likeness of hypotheses or theories could be determined (Ojong 234). Corroboration simply means for Popper the degree by which a theory is falsifiable (Uduigwomen 234). Thus, if a theory is put to test, and it passes such test, then the theory must be corroborated according to Popper, and if the theory fails the test, it is not corroborated; hence it should be jettisoned or rejected. From the above exposition of Karl Popper, it could be seen that his ideas on induction, falsifiability, verisimilitude and corroboration and many others were indeed of positive influence to the growth and development of rationalism in modern science.

Imre Lakatos Concept of Scientific Research Programme and its impact on Modern science

Imre Lakatos was born on November 9, 1922 in a small town called Debrecen in Hungary and died on 2 February, 1974 in London. He specialized in philosophy of science, philosophy of mathematics and epistemology. His notion of science was greatly influenced by Karl Popper's work on critical rationalism, this is why Newton-Smith describe Lakatos as "the revisionary Popperian" (Uduigwomen 237). Lakatos made a remarkable improvement on rationalism in modern science through the blueprint of his predecessor Karl Popper and further attempted to correct the deficiency in Popper's philosophy of science (Newton-Smith 77). Thus, Popper had admitted that the growth of science is possible when a theory is put to the test called the falsifiability test. Any theory that stands the success of scientific test should be corroborated or jettisoned. Popper's view on how a theory should be jettisoned was heavily criticised by scientists and philosophers of science. But Imre Lakatos took a new dimension to overcome the criticisms against Popper by attempting to improve on Popper's limitations. In his work titled "The Methodology of Scientific Research Programme". Lakatos expressed his thought on what could be the proper method for science. He reasoned that science grows from a sequel of theories each one generated by modifying the one proceeding it (Uduigwomen 237). Lakatos believed that science does not grow by jettisoning the anomalies. To overcome this challenge he modelled a procedure that allowed anomalies theories to be re-examined rather than discarded (Newton-Smith 78).

However he presented a picture of new scientific paradigm that has its development through the process of critical evaluation of theories and stated that the generation of anomalies is not a sufficient condition for rejecting a particular theory, because for him, a theory with anomalies is better than no theory at all (Newton-Smith 78). All these ideas were developed in his *Scientific Research Programme* in attempt to improve and overcome the objections in Popperian falsificationism (Chalmers 80). In his Research Programme, Lakatos explicates two fundamental concepts that constituted the major components of his programme thus:

1. The positive heuristic
2. The negative heuristic

The positive heuristic according to Lakatos consists of rough guidelines, hints, or suggestions indicating how the research programme might be developed or changed and how to modify and sophisticate the protective belt (Ojong 71). The negative heuristics involves the basic assumptions underlying the programme which can be seen as the basic theoretical postulate or axioms of a theory (Newton-Smith 82). The research programme is characterised by a hard-core of theoretical laws which are defended and protected from refutations by a network of auxiliary hypotheses, they can turn the anomalies into positive evidence, and can contradict any rival theories in other to protect a theory from being jettisoned (Aigbodioh 65).

An examination of Lakatos research programme shows the positive contribution to modern science. Thus most theories in science are not jettisoned because they do not solve current scientific issues but are preserved because they remain an asset to the development of other scientific theories in modern science, an aspect that needs intelligibility of science. Since the time of the development of Lakatos methodology in science, his ideas have become an asset to the movement of rationalism in modern science. And up-till today, the entire gamuts of his philosophy have strengthened the modern rationalists' moment.

Gottfried Leibniz Concept of Monads and its impacts on Modern Science

Gottfried Leibniz is another modern rationalist philosopher who took delight in arguing on the utility of rationalism in knowledge acquisition especially in science and philosophy. He was born on July 1, 1646; at the age of 15, he started his university education at the city of Leipzig. He specialised on mathematics, law, and philosophy of science. Leibniz went into a philosophical investigation to unravel the nature, scope and property of the natural universe. It was one of his topical desires to have a deep understanding of the mechanism of the universe. This quest about reality influenced him to seek for rational explanation about the universe. The quest for his rational investigation of the universe was the quest to seek answers to philosophical questions within his era. Such as, what is the origin of things in the universe? Where are we coming from? Where are we going to? And from where do men originate? These are some of the thought provoking questions Leibniz sought to investigate. He drafted a mechanism to unravel the mysteries behind the things that exist in the universe and finally argued that the fundamental unit of every reality is known as Monads.

The term Monads is described by William Lawhead as a concept derived from a Greek word 'Monas', meaning unity or 'that which is one'. According to Leibniz the entire universe is made up of infinity of simple, nonmaterial, invisible substance called monads (*The Voyage of Discovery* 265). To articulate this notion, Leibniz distinguished two kinds of truth: truth of reasoning and truth of fact (Bryan Magee 97). The truth that dwells on the existence of monads can only be derived from reasoning, and when monads comes together to form macro element at this stage they can be seen as truth of fact. Thus, the foundation of truth comes through reasoning as subscribed by this work.

In the work of Samuel Enoch Stumpf, he describes the properties of monads as being un-extended, having no shape or size, it is a point, not a mathematical or physical point but a metaphysical existent point, each monad does not depend on the other to exist, and monads do not have any causal relation to each other (243). Leibniz described that monads follow its own purpose and form a unity of orderly universe. William Lawhead further describes the monads as possessing the quality of self-enclosed and self-sufficient unit which means that nothing enters and nothing goes out of a monad. This is what Leibniz described as the windowless monads (265).

The implication of this concept in modern science is based on the idea that reality should not be seen as a physical entity alone, this ideology is what possibly influenced the modern understanding of atoms. Today, the idea of atom has further developed to encompass various units or subatomic element like protons, electrons, photons, quarks, mesons and neutron. It is an area that is today studied under quantum mechanics (the study of subatomic particles) in physics. Thus, the growth of rationalism in Leibniz philosophy created more understanding on the aid of reason in understanding the concept of monads and also gave science the theoretical background to the study of unseen reality that can only be understood through the tool of reason as studied under quantum theory in modern science. His distinction on the two kinds of truth: truth of reasoning first and truth of facts as it connects to the study of monads is a critical support for the thesis of this work. it supports the notion that rationalism is the foundation of modern science.

Isaac Newton's Impact on Rationalism in Modern Science

Isaac Newton is another great scholar that gave meaning to science and created a science that is justified by reason. Modern Science was totally influenced by the works of Isaac Newton. He was an intelligent scholar and indeed a chief advocator of modern systematic thinking. The era of Isaac Newton was remarkable and fashioned in objectivity, causality and determinism to all scientific knowledge. These three assumptions in modern science help to systematize science and compelled scientists to abide by a remarkable standard all over the entire community of science. Isaac Newton reduced science into precise mathematical laws capable of pushing in objectivity and shaping the quest for scientific knowledge with the aid of reason. His laws explain everything about the world of matter (Macro world) and everything ever seen using sense perception were all explained by the laws of Newton. These laws are what Newton called the three Laws of motion and the universal law of gravity.

Isaac Newton revolutionized physics and science with his postulation that all bodies are governed by the three laws of motion which he used in describing the motion of the planets and the moon. Let us consider his three laws as posited by John Gordon (et al):

1. The first law of motion (also called the **law of inertia**) states that an object at rest will continue to be at rest and an object in motion will remain in motion with a constant velocity unless acted on by a **net external force.**

2. The second law (also called **the law of acceleration**) states that the force applied to an object is proportion to its mass multiplied by acceleration (F = Ma). Meaning the acceleration of an object is directly proportional to the net force acting on it and inversely proportional to its mass.

3. The third law (also called **law of action-reaction**) states that for every action there is an equal opposite reaction. (*Principles of Physics* 64)

Thus, with these three laws, Newton created a whole new model of the universe, superseding Ptolemy's model of epicycles; eighty years before Galileo had pointed out that the earth rotates around the sun, and also brought in a mechanistic view about the universe. This Mechanistic view about the universe by Newton and Galileo provided the basic for 17th to 19th century cosmology. Mechanistic view sees the universe as an arrangement in which stars and planets interact with each other like springs and cogs in clockwork. Thus, if the initial positions and states of all objects in a mechanically determined universe are known, all events can be predicted until the end of time, simply by applying the laws of Newton, no observation will be required but by mere reference to reason. The mechanistic view of the universe brought about a new hypothesis or method of interpreting reality using determinism. Underlying determinism is the view that everything in the universe, including all the motions, from the smallest to the largest occur in a way that can be predicted with absolute accuracy using the laws of Newton, nothing is left to change (Pagels 4). It is quite imperative that one cannot discus rationalism in modern science without having a critical survey of the contribution of Isaac Newton and Galileo because they were among the greatest figures that shaped science in the modern era. This is why many scholars regard Isaac Newton as the father of modern science.

In line with Pagel's exposition, Akpan on *Quantum Mechanics and the question of Determinism in Science* contends that:

> Classical science and in fact Post-Newtonian science up till the early twentieth century were mired in a deterministic interpretation of realities. The deterministic hypothesis in science holds that everything in nature has a cause and if one could know the antecedent causes, he could predict the future with certainty (*Sophia: A Journal of Philosophy* 72).

The above explains that in the classical physics, determinism and prediction were all possible. Gary Zukav, on *The Dancing Wuli Masters: An Overview of the New Physics* rightly agreed on the predictive powers of classical mechanics and opines that:

> Newton's laws of motion describe what happened to a moving object. Once we know the laws of motion we can

predict the future of a moving object provided that we know certain things about it initially. The more initial information that we have, the more accurate our predictions will be... for example, if we know the present position and velocity of the earth, the moon, and the sun, we can predict where the earth will be in relation to the moon and the sun at any particular time in the future giving us a foreknowledge of eclipses, seasons, and so on (50).

He further states that:

According to the old physics, however, it is possible, in principle, to predict exactly how a given event is going to unfold, if we have enough information about it.... The ability to predict the future based on knowledge of the present and the laws of motion gave our ancestors a power they had never known (51).

Gary Zukav's work gives us the background of the predictive power of classical mechanics using Newton's laws full of certainty. The implication of Isaac Newton's contributions to modern science is shown by its insightful explanation of the nature of the universe by the aid of reason. Through Newton rational understanding of nature, the entire world of science was developed and there was a strong revolution in science that shaped the entire scientific world into modern cosmology.

CHAPTER FIVE

ISAAC NEWTON AS A RATIONALIST: THE BIRTH OF NEWTONIAN MECHANICS

Introduction

Sir Isaac Newton is considered by many as a magician and one of the greatest scientists of all time. He is a key personality in the history of scientific revolution. He brought in systematicity to science, and created a platform for scientific knowledge based on observation, experiment and testing of facts, an approach which was not common in science. He conceptualized the thesis of Galileo and energises science into modern light. Before now, science did not attain systematic approach, as Mysticism, Sorcery, Voodooism, Magic, Wizardry, Diablerie, Mojo, Thaumaturgy, Necromancy and many others were all seen to be scientific. Thus, the idea of observation, evaluation and testing of facts came into science as a result of Galileo and Newton's efforts to picture the entire universe mechanistically. Isaac Newton through observation and experimentation postulated laws of mechanics that overturned the long standing ideas of Aristotle which had lasted for decades in science. He gave a sharp distinction among space, time, speed, motion, gravity, light etc. All these, he assigned fundamental unique properties, that control the entire forces of the universe using his laws.

His laws of mechanics introduced determinism to science and gave scientist the power to predict the future with great certainty; meaning through the study of a cause, the effect can be known. And through the observational effect, samples can be gathered, by which an induction is made. This induction is what possibly gave rise to the formation of theories which act as the covering law for science in the formation of further predictions.

The philosophical implications of this tradition led to the formation of empiricism in science, a traditional belief in philosophy that claims reliable scientific knowledge through sense data collection using the five sense organs, sight, hearing, smelling, feeling, and tasting, and through experiment and testing of fact evidence can be achieved. This method guided Newton to give an explicit account of the nature of light, the movement of large bodies (motion) and the state of gravity. Thus, he is a rational genius and a very complex and secretive man that up till now no one can understand the power of his creative thought and the origin of his thought-provoking ideas. He calculated the motion of bodies and revealed that nature can be reduced to a precise mathematical formula, an area in mathematics today known as *Calculus*. He also studied the nature of light, stating all the distinctive properties of light using a prism called Newton's theory of light. Through observation, he postulated the laws of inertia and proved the properties of matter by which objects continue in their existing state or uniform motion in a straight line, unless changed by external force.

The search for the true nature, scope, and origin of the universe was Newton's greatest dream in his entire life. He relied on no one, but himself, he has a fertile mind always ready to accept correction through criticisms. He used natural philosophy to uncover the laws of the universe which led him to certainty in knowledge. By 1664 at the age of 21, he started being troubled by certain philosophical questions, what he called philosophical issues. These issues included time, eternity, comets, atoms, attraction, sun, plant, air, meteors, electrical, vision, colour, light, gravity, vacuum, density, heat and cold, etc. For Newton, these philosophical issues would give him a clear picture of the entire mechanism of the universe. His formulation of the three laws of motion placed the science of mathematics on a solid foundation. By inventing the calculus, Newton gave physical science a new powerful kind of mathematics that is still in vogue in today's study of the natural universe.

In summation, this work is a philosophical investigation of Isaac Newton's discoveries in science and their philosophical implications. We posit that, the philosophical implication of Newton's revolutionary science is one that should not be forgotten in a hurry. This work captures such themes as brief biographical sketch of Newton, specific discoveries of Newton and their philosophical implications and so on.

Biographical Sketch of Isaac Newton

Newton was born on 25 December, 1642 in Woolsthorpe, Lincolnshire, England. His mother was Hannah Newton (nee Ayscough). Newton was her first child. She named the boy Isaac in honour of his father, a farmer who passed on two months earlier at the age of thirty six. Baby Newton was born premature, making his mother to presume that he might not live out his first day. As a child he was not happy because he lost the comfort of living with his biological father due to death. When he was three years, his mother married another man Barnabas Smith, a minister twice her age. At this point Newton was sent to live with his maternal grandmother doing farm work together with his studies. He was separated from his mother for nine years until the death of his step-father in 1653. Because of his experience with women, he had little or no time with women until his death he never married, but focused attention on his career. Newton did not have any hope of becoming a legend in science, because of his poor background and the condition he found himself as a poor boy in the farm with the grandmother. The nine years he spent in Woolsthorpe away from his mother was indeed a painful moment for young Newton, a time until his death was a sobering moment in his life time. In his book *A Portrait of Isaac Newton*, Professor Frank Manuel concluded that the remarriage of his mother was the most critical episode in Newton's entire life (Brennan, 15). This is giving us a strong feeling that Newton never agreed to the remarrying of his mother. He started his career with his famous book *Philosophiae Naturalis Principia Mathematica* (Mathematical Principles of Natural Philosophy) published in 1687. This book astounded the world scholarship and introduced into science a systematic approach of understanding the covering mysteries of the entire universe. With this book, Newton solved the greatest problem in the history of science up to that time, the problem of the mechanics of the universe (Brennan, 12). Newton was actually a complex man; his secretive style would baffle any scientist as to how he got through in his discoveries. When he was at Cambridge, he taught mathematics in Trinity College where he went into complex mathematics that gave way to a new set of mathematical system now called *Calculus.* Before then, he informed no body until when the real success was achieved. This is why Stephen Hawking, in his best-selling book *A Brief History of Time* admitted that "*Newton was not a pleasant man. His relationship with other academics was notorious, with most of his later life spent embroiled in heated disputes*" *(191).* He was once appointed the President of the Royal Society of London for Improving Natural Knowledge commonly known as the Royal Society, and also served the British government as Warden and Master of the Royal Mint. Newton built the first practical reflecting telescope and developed a theory of colour based on the observation that a prism decomposes white light into the many colours of the visible spectrum. He formulated an empirical law of cooling, studied the speed of sound, and introduced the notion of a Newtonian fluid. In addition to his work on calculus, as a mathematician, Newton contributed to the study of power series, generalised the binomial theorem to non-integer exponents, developed a method for approximating the roots of a function, and classified most of the cubic plane curves.

Newton was a Fellow of Trinity College and the second Lucasian Professor of Mathematics at the University of

Cambridge. He was a devout but unorthodox Christian and, unusually for a member of the Cambridge faculty of the day, he refused to take holy orders in the Church of England, perhaps because he privately rejected the doctrine of the Trinity. He died on 20th March, 1726.

Newton's Contributions to Science

In modern science, the ideas and discoveries of Newton are regarded by many as one which have brought in a great revolution to the entire scientific community. His ideas span through physics, chemistry, astronomy, religion and numerous disciplines. He was a master of his own. He believed that nature is controlled by certain absolute force, and that the universe exists because of the agreement between its components and the forces on it. This force, according to Newton, is what gives rise to the way objects behave; he called this the force of gravity. Let us examine most of Newton's contributions to science.

Isaac Newton on Gravitation

In 1679, Newton returned to his work on (celestial) mechanics by considering gravitation and its effect on the orbits of planets with reference to Kepler's laws of planetary motion. In physics, the term gravitation is quite synonymous with gravity. Gravity is a scientific concept used in describing the force that attracts a body towards the centre of the earth or towards any other physical objects with mass. This force can also be called gravitational force; it is the mutual force of attraction between any two objects in the universe; thus the weakest of the fundamental forces (Serway, 83). However, according to Newton, every object with mass is affected by this force. It is the universal force that controls all the properties and behaviours of large scale bodies. This force is called the gravitational force. It is described by many as a single magical force that runs across the universe. Newton combines mathematics and mysticism to uncover the force that controls the entire universe. The force of gravity is seen as a vital agent that holds and controls the entire Milky Way, solar energy, and causes the dropping of rain, the positioning of the moon, the rotation of earth around the sun, and also the existence of million stars in the galaxy. By further application, it is the agent that gives weight to objects with mass and causes them to fall to the ground when dropped. We have four fundamental interactions of nature; the first is Gravitational force (attractive force between objects due to their masses), Electromagnetic force (between electric charges at rest or in motion), Strong nuclear forces (between subatomic particles), and Weak nuclear force (accompanying the process of radioactive decay) (Serway 63; Igwe 350). Thus, gravity is what causes the earth to exist and pivot around the sun, the sun, moon and stars and most macroscopic objects that exist in space. The moon stays around the earth because it is held by gravity, and thus the entire mechanism of the universe is propelled by the universal law of gravity.

Modern work on gravitational theory began with the work of Galileo Galilee in the late 16th and early 17th centuries in his famous experiment of dropping of heavy and lighter balls from the Tower of Pisa. This experiment shows that gravitation accelerates all objects at the same rate, thus refuting the Aristotelian theory that heavier bodies fall faster than lighter ones. By implication, all free falling objects will fall towards the earth at the same time irrespective of their weight (Igwe 351). This Galilean notion influenced Newton to see the entire universe as the handwork massive clock built by a complete mechanism called gravity. All celestial bodies obey this law through the force of attraction that exists between two objects with masses.

Isaac Newton through physical observation of the nature of the universe discovered that objects are attracted to one another in so far as these objects possess some element of masses. Thus, these objects obey certain natural law called the law of gravity. It is natural, because it is not created by man and thus states that the force of attraction between two objects is directly proportional to the product of their masses and inversely proportional to the square of the distance between them. Newton showed that the same laws of nature are applicable to every object with masses; the gravitational force that attracts apples to the ground is identical with the gravitational force that keeps the planets in their orbits; all are guided by the same law.

In the law of gravitation, Newton found the solution to the problem of planetary motion and gave science a powerful tool for understanding natural phenomena, explaining that it is only by gravitational pull that the planets and sun are held together as shown in the diagram bellow.

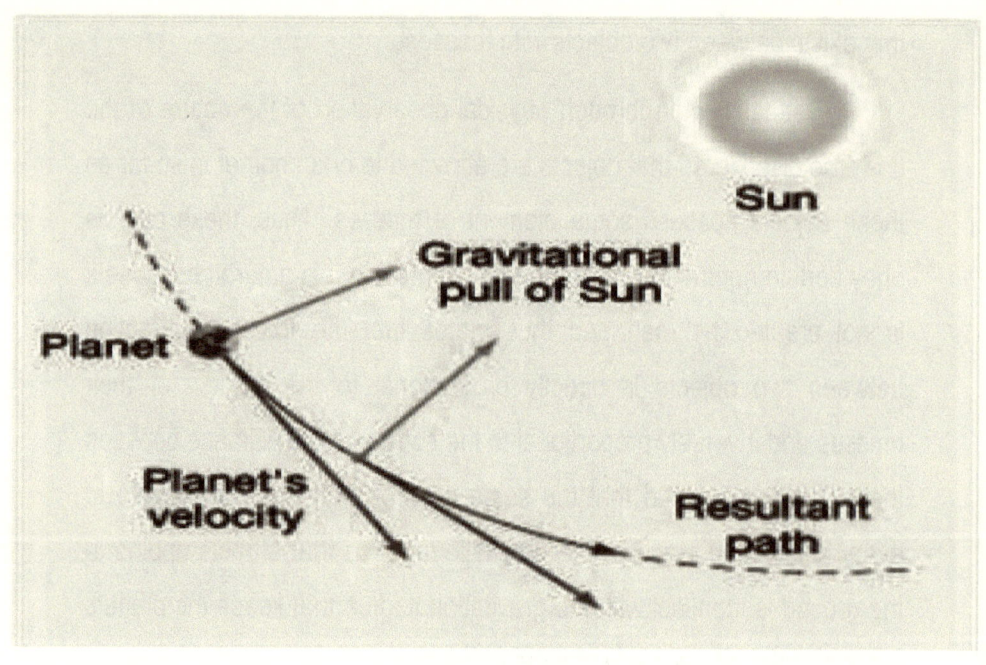

Through this understanding, Newton realized that to make a planet take an ecliptical or circular orbit required an attractive force between the planet and the sun, he then formulated the law of universal gravitation, stating that the attractive force between any two bodies is proportional to

$$F = G\frac{m_1 m_2}{r^2}$$

the products of their masses (kg) divided by the square of the distance between them. This can be described symbolically as:

The above formula is one of Newton's greatest equations in 20th century Physics that gave more strength to determinism in science. Where F= force, G=gravity, m_1= mass of object one and m_2= the mass of object two and r^2 = the square of the distance. When you have the same figures at various points, it is possible to achieve certainty in your findings anytime, and anywhere you ever want your findings. Thus, this is what gives rise to objectivism in science.

The Notion of Motion in Newton's Cosmology

The idea of gravity extensively deals with the movement of planetary bodies around its orbits through what is described in modern science as motion. Motion can be defined as the displacement of body from one point to another in space and time. Therefore, a free falling body moving freely under the influence of gravity must participate in some actions of motion. In other words, when a body or an object changes its position with respect to time and it goes on continuously, we say that the object moves. This movement is what can be described as object in motion.

Before Newton the idea of motion was not new to science. Aristotle had earlier articulated his idea of motion of bodies. For Aristotle, everywhere we find evidence of design and rational plan. No doctrine of physics can ignore the fundamental notions of motion, space, and time. Motion is the passage of matter into form, and it is of four kinds:

(1) Motion which affects the substance of a thing, particularly its beginning and its ending;

(2) Motion which brings about changes in quality;

(3) Motion which brings about changes in quantity, by increasing it and decreasing it; and

(4) Motion which brings about locomotion, or change of place. Of these the last is the most fundamental and important (Cited in http://www.iep.utm.edu/aristotl/).

` This notion of motion by Aristotle laid the foundation for modern theory of motion, and therefore led Newton to develop a new set of rules that control all bodies in motion. This he called the laws of motion.

Newton on Laws of Motion

Newton through observation gave a distinctive interpretation of every object that moves within the natural universe as that which obeys what he called the laws of motion. These laws serve as an operational criterion for every object in motion or at state of rest. The ways objects behave in space, on the sea, land and in the air are all connected to the laws of motion. He conceptualized these laws into three; and these include:

1. First Law of Motion (Law of Inertia)

Newton's first law also called the law of Inertia, states that an object at rest will remain at rest and an object in motion will remain in motion with a constant velocity unless acted on by a net external force.

Newton's first law effectively says that the natural motion of an object at rest remains at rest, or, if it is moving, moves in a straight line at constant speed. In other words, if there are no forces acting on an object it will either remain at rest or move with constant speed in a straight line. According to Newton, it takes only an external force to change either the speed or the direction of motion. Philosophically, if we agree on Newton's first law, what will be the consequences of the moon in space, if there was no force on it? This may inform us that over there in the space, the movement of the planets, the trajectory of the moon around the sun is held by the force of gravity, proving the validity of Newton's first law. Thus, the result of this law is that, in all acceleration there is a force.

2. Newton's Second Law (Law of Acceleration)

Newton's second law, the law of acceleration, states that the acceleration of an object is directly proportional to the net force acting on it and inversely proportional to its mass. Thus, the direction of the acceleration is the direction of the net force.

Force change in quantity of motion per unit time. If mass is constant, then this becomes:

Force (mass) x (change in velocity per unit time), and since

Acceleration = (change in velocity)/time,

$F = (mass) \times (acceleration)$.

$F = ma$.

Where F= force, m = mass, and a = acceleration.

The philosophical implication of this second law is that, when different forces act upon the same mass, the greater force produces the greater acceleration, and when the same force acts upon different masses, the greater mass receives the smaller acceleration, thus the acceleration will occur in the same direction as force.

3.   Newton's Third Law (Law of Action-Reaction)

Newton's third law, the law of action-reaction, states that when two bodies interact, the force which body "1" exerts on body "2" (the action force) is equal in magnitude and opposite in direction to the force which body "2" exerts on body "1" (the reaction force) (Serway, 64).

The implication of Newton's third law is that, when we push a car to move upward, there is always a reacting force from the car stubbornly resisting smooth movement of the car. Thus, this reacting force from the car sends back a reaction exerting a force on the pusher. In other words, we cannot exert a force on the car without it exerting a force on us. Because of this experience, Newton came up with his third law stating that for every force there is an equal and opposite reaction, meaning that whenever one object exerts a force on a second object, the second exerts an equal force in the opposite direction on the first. Also, when a man pushes backward on the earth with his foot, the earth pushes forward on him. It is the later reaction force that leads to his forward motion.

Isaac Newton on Optics

On the question, does light really exist? What is the nature and property of light? To solve this problem in physics was one of Newton's great dreams. He so much believed that without proper investigation on the nature of light, man would continue to operate with vagueness and certainty in science will a great failure.

In 1666, Newton carried out an experiment on light using a prism (In Optics, a prism is a triangular transparent object used to separate white light to spectrum of colours). Thus, he observed that the spectrum of colours exiting a prism in the position of minimum deviation is oblong, even when the light ray entering the prism is circular, which is to say, the prism refracts different colours by different angles. This led him to conclude that colour is a property intrinsic to light, a point which had been debated in prior years. Thus, the outcome of Newton's experiment on light shows that, the light we receive from the sun or from an incandescent lamp is usually considered to be white light. But this light is, according to Newton, composed of many colours. We can observe the breakup of white light into its component colours by passing the light though a triangular piece of glass called prism (Marion,420).

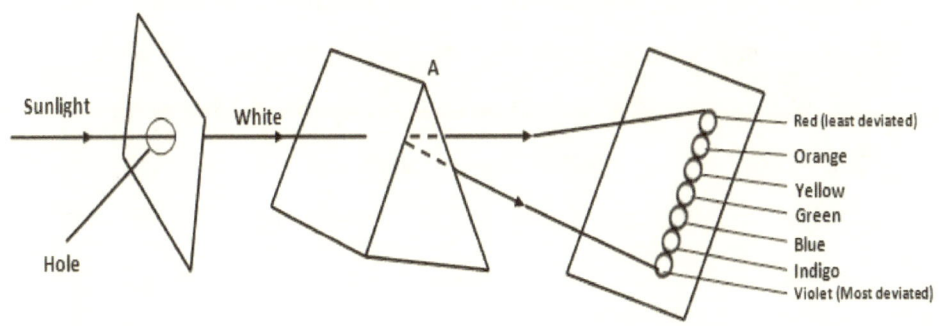

The diagram above shows a schematic representation of white light, consisting of all colours, passing through a triangular glass prism, the white light is dispersed into a spectrum of colours. The red light refracted least and appears on the left-hand side of the spectrum, followed by orange, yellow, green, blue and violet. The reason that light disperses into different colours is because light is can act as a wave at a time and can also act as a particle in another point.

From 1670 to 1672, Newton had lectured on optics and during this period he investigated the refraction of light, demonstrating that the multicoloured spectrum produced by a prism could be recomposed into white light by a lens and a second prism. Modern scholarship has revealed that Newton's analysis and re-synthesis of white light owes a debt to corpuscular alchemy. He also showed that coloured light does not change its properties by separating out a coloured beam and shining it on various objects. Newton noted that regardless of whether it was reflected, scattered, or transmitted, it remained the same colour. Thus, he observed that colour is the result of objects interacting with already-coloured light rather than objects generating the coloured themselves. In other words, the colours that object posses exist extrinsically by the object which exists intrinsically by light. This is known as Newton's theory of colour.

Philosophical Implications of Newton's Contributions to Science

1. On Determinism and Causality

Determinism is a philosophical doctrine that asserts that all events and action are determined by causes. And in causality, in every cause there is an effect. The effect must necessitate from the cause. In other words, nothing happens by chance, the universe that rotates the shining nature of stars, the heart of the sun, the cold nature of ice, dropping of water from the sky and the break and fall of the night are all events that are covered within the laws of determinism and causality. Classical mechanics, which is known by many as the physics of Isaac Newton has the embodiment of all laws of the physical universe. Its studies the laws guiding the behaviour of large bodies and the way these bodies interact with other objects, the space they occupy and the limit of time they can move from one place to another; all these could be achieved with certainty. In other words, Newton brought certainty to science as against the then speculative approach of understanding the universe. Newton through his laws of motion and universal gravity asserts that, an object can cause another to behave in a different direction in accordance to the law. Through observation and experimentation it was possible for the cause and effect to be determined, which gave power for prediction to be made, and provided science the ability to predict and determine future occurrences with great certainty.

Determinism in classical mechanics was possible because of the great postulations of the greatest genius in science called Isaac Newton. Many may ask, how? Newton through experiment and observation of the entire universe discovered the magic of how large bodies behave with one another from time to time. Through this he came out with the universal laws guiding the movement of all seen bodies, the way they react with one another. These laws shaped the way he understood the universe and the way scientists make predictions.

2. On Empiricism

Empiricism is the belief in sense perception, induction, and, that, there are no innate ideas. G.O. Ozumba in his work entitled "*Isms in Philosophy*" stated that empiricism is a "School of thought that holds that knowledge is got through sense experience. The five senses of hearing, seeing, smelling, tasting and feeling are five important ways of getting acquainted with the external world" (*A Concise Introduction to Philosophy and Logic* 47). He further defines empiricism in his *Concise Introduction to Epistemology* as an epistemological school that based knowledge on experience, observation or on experiment rather than theory. Like any other school of philosophy, empiricism is of the view that, the only means for certainty in knowledge can be derived from observational factors or facts that are capable of being experienced. Isaac Newton's philosophical search for the substance of the universe can be seen as the powerhouse of empiricism. He never believed in what cannot be practically experienced. His works on optics was achieved through careful experimentation of light to capture the nature of light. His laws of motion and universal laws of gravity were all empirically based and thus, strengthened the place of sense data collection in science. Though a Christian, he doubted the existence of trinity (the belief of three in one God), which was then seen as heresy by the church.

Limitations of Newtonian Science

One of the greatest defects of Newton's science is that it is only operational at the level of the macro world. In Physics, his laws are only applicable to seen objects, that is, objects "out there". This is why it was possible to apply Newton's law in solving the problems of the rotation of the earth, the shining stars, moon, sun, etc. Thus, Igwe admitted that Newton's laws are applicable to objects of our everyday experience to solve our problems (355).

The limitation of Newtonian laws is the inability to penetrate into micro world, the world of the very small, and the world of sub-microscopic reality. Microscopic world is the realm studied by Quantum theory alone. The laws of Newton cease to exist at the realm of quantum mechanics because such realm is not penetratable using human senses. Our eyes cannot see quantum events, or observe the nature of photons or electrons, we cannot hear the sound of Plasmas neither can we torch an atom. It can only be studied using the foundation of quantum theory; the only existing theory that studies spins, quarks, mesons, electrons, etc. To this effect, Newtonian laws only study large objects under classical mechanics, while Quantum theory studies small objects under Quantum mechanics.

Conclusion

Isaac Newton is regarded by many as one of the greatest scientists in 20th century. He clearly reveals the physical universe as a clock work mechanism, with a set of fixed knowable laws, the laws of motion and the universal law of gravitation. This background and fixed attitude influenced a lot of things in his career and gave rise to his notion in mathematics called *Calculus.* He reduced all forms of natural phenomenon, the world of matter, and all abstract entities into a precise mathematical code called the calculus. This mathematical code models modern scientists to apply exactitudes in all scientific findings. But the most challenging aspect of calculus was that it was too difficult to understand, and up till today, very many still find it difficult to understand. All his ideas helped to shape modern science, and philosophically they paved way for certainty in science. Certainty here implies that, with the application of the laws of Newton, one can arrive at a universal objectivity in all findings. Newton's theory enjoyed its greatest success when it was applied to predict the existence of another planet called the Neptune. This was possible based on the motion of other existing planets, thus paving way for more planets to be discovered.

In today's Physics, the most challenging moment of Newton's mechanistic understanding of the universe is that, his theories do not work when applied to objects moving at the speed of light. It also ceases to work when subjected to the micro-world of matter. It is only operational at the macro world, the world at the very large.

Because of some these inadequacies, quantum mechanics has come to shatter most of the fundamental laws of Newton to explicitly explain the nature of subatomic particles like quarks, photons, plasmas, mesons, electron, neutrons, and many others. In any case the laws of Newton have not failed scientists when dealing with objects at the very large, their predictive powers are still applicable at all moving large objects, and gravity is still operational because objects are still in a constant interaction with one another.

Finally, many would admit that despite some of the philosophical limitations of Newton's discoveries in science, he is still a great scientist when it comes to his enormous contributions to science.

CHAPTER SIX

THE PHILOSOPHY OF ALBERT EINSTEIN
E=MC²

Introduction

$E=mc^2$ to so many is Albert Einstein's biggest idea and also regarded as one of the world most famous equation ever discovered about 115 years ago. This equation excited physicist and showed that mass and energy are not distinct, but are two forms of same thing. E for Energy; M for Mass; C for the Speed of Light, and 2 for "Squared". Albert Einstein is considered by many to be the greatest genius of all times. His virtues proved the same and his discoveries might well pass for miracles. Many recognize him in Europe as a supper-human for giving science a new dimension on how to understand the universe and reality as a whole. Einstein reviewed the nature of the universe allowing us to understand that reality can be reduced to a precise mathematical formula, $E=mc^2$. This hidden mystery about matter, energy and light helps to unite all reality into one universal framework. This paper is an insightful exploration of and a deep reflection on the fundamental issues connected to this famous formula. Also, it is an attempt to insightfully investigate the impact and philosophical criticisms against Einstein's dream regarding his famous. Philosophically, this work is significant because $E=mc^2$ is a refinement of everyday thinking. That is why we shall examine the equation from the perspective of the various traditional branches of philosophy.

In every discipline, there must be a philosophy. This is because, to many, philosophy is considered as the Mother of all disciplines. Hence, the ability of philosophers to explore all disciplines is regarded as plausible and important in the growth and development of knowledge. In other words, philosophy by its natural, epistemic and investigative ability does not have boundaries. This is why we have the various "philosophies of". Example, philosophy of Law (Jurisprudence), Philosophy of Chemistry, Philosophy of Physics, Philosophy of Social Sciences, Philosophy of Economics, Philosophy of Science, Philosophy of Education, etc. In all this, it is the application of philosophical tools to the assumptions and presuppositions of these disciplines. Thus, the tools of criticality, analyticity, evaluation, and logicality are often considered.

However, the work of a great physicist called Albert Einstein is considered as one of the most important and monumental revolutions in Philosophy of science. Various thinkers in philosophy and physics consider his work as the greatest revolution which eventually won him a Nobel Prize in physics. With great power of thinking, after proper investigation of the physical universe, he considered E (Energy) and M (Mass) as one united entity at C (speed of Light). As we shall see this equation as incomplete, as its lacks the fundamental basics in capturing all reality, especially at the level of subatomic particles.

A Brief Biographical Sketch of Albert Einstein

Albert Einstein was born in Ulm on March 14, 1879 to Hermann and Pauline, an entirely irreligious (Jewish) middle-class German parent. He was always curious to understand the things around his environment as a child. He is well known for his special and general theory of relativity. He spent his early life in Munich, where his family owned a small electrical shop. In 1905, he received his Doctorate degree from the University of Zurich for a theoretical dissertation on the dimensions of molecules. The equation $E=mc^2$ is a famous equation attributed to Einstein which is the central theme this work aims to examine. Before we proceed, let us give a background of the nature of energy and mass during Newtonian physics.

The Nature of Energy and Mass in Newtonian Physics

The idea behind the equation has a long history, starting with Heraclitus 535 BC, Democritus 460 BC, Aristotle 384 BC, and Lucretius 99 BC. Newton's work and the subsequent development of classical mechanics showed that matter and the motion of matter somehow were related. Hegel's dictum (Just as there can be no motion without matter, so there can be no matter without motion) was the foundation of classical mechanics (Hegel 197).

Newtonian physics was an era that was dominated by the physics and/or ideas of a genius called Isaac Newton. His dominance in physics or science was as a result of his curiosity to understand the composition of the universe. This helped him to establish fundamental laws that helped in piloting and shaping all scientific discoveries in his era. Examples are the laws of motion, law of conservation of mass, and law of gravity, etc. Isaac Newton discovered that Nature has its three unique properties, which control, direct, and give birth to the entire universe. These include Time, Space and Matter. He posited that these three unique properties give every reality its forms and shapes. Thus, they define the existence of bodies, events and all occurrences and even the way we see objects. They were fundamental properties of nature that give macroscopic picture of the universe. In other words, Newtonian Physicists study the macro world.

However, every discovery during Newtonian age must fit into the above three properties of nature. Hence scientific discoveries were limited to space, time and matter. Thus, Energy and Mass were what gave Time, Space and Matter their unique existence. This is why it was believed that in every macroscopic element, there must be an energy or mass associated to it. Energy here means the property of matter and radiation which is manifest as a capacity to perform work. And mass refers to the quantity of matter which a body contains, as measured by its acceleration under a given force or by the force exerted on it by a gravitational field. (Concise Oxford English Dictionary 906) Thus, Mass and were seen as a unique but separate properties of matter, space and time. Nothing was seen as a connecting point between them. But after the formation of the theory of relativity, Albert Einstein did not only unite space and time to Space-Time, but provided science with a unique equation $E=mc^2$, which was capable of solving so many scientific problems, meaning Energy and Mass are two interchangeable and/or inseparable entities. This became an outstanding revolution of mass and energy relations in science.

The Origin and Interpretation of $E=MC^2$

Mass–energy equivalence arose originally from special relativity as a paradox described by Henri Poincaré. Einstein proposed it in 1905, in the paper *Does the inertia of a body depend upon its energy-content?*, one of his *Annus Mirabilis (Miraculous Year) papers*. Einstein was the first to propose that the equivalence of mass and energy is a general principle and a consequence of the symmetries of space and time. (Cited in https://en.wikipedia.org/wiki/Mass%E2%80%93energy_equivalence)

Because it is the most confused equation in science, Students and some scientists find it difficult to understand the equation $E=MC^2$. $E=mc^2$ stands for: E (Energy of a physical system), M (Mass of the system) modern concept sees M as rest mass, and C (speed of Light in a vacuum which is about 3×10^8 m/s). In line with this, Corkcroft and Walton opine thus:

> It followed from the special theory of relativity that mass and energy are both but different manifestations of the same thing -- a somewhat unfamiliar conception for the average mind. Furthermore, the equation E is equal to m c-squared, in which energy is put equal to mass, multiplied by the square of the velocity of light, showed that very small amounts of mass may be converted into a very large amount of energy and vice versa. The mass and energy were in fact equivalent, according to the formula (Cited in http://zebu.uoregon.edu/2004/hum399/lec18.html)

Thus, let us quotes Wong Chee Leong & Yap Kueh Chin at long length on a schematic representation of the development in Einstein $E=mc^2$ during its 100 years, summarized as follows: this is elaborated by Leong & Chin in the following schemata.

1. Before 1905 (The forthcoming of E = mc2)

1881	J. J. Thompson proposed that a charged conductor in motion increases its mass by (4/15)µe2/a
1904	H. A. Lorentz proposed that mL = m0(1 – v2/c2)-3/2 based on deformable spherical charge.
1904	Hasenöhrl derived an apparent mass increase of a moving cavity containing electromagnetic energy E, obtaining δme = 8/3 E/c2.
1904	On Abraham's suggestion, Hasenöhrl corrected this to δme = 4/3 E/c2

2. From 1905 to 1924 (The beginning of E = mc2)

1905	Einstein proposed that mL = m0(1 – v2/c2)-3/2 and E = mc2. A body's mass diminishes by L/c2 if it gives off energy L.
1907	Einstein formulated rest energy explicitly. Energy and mass are equivalent. E= m0c2 (1 – v2/c2)-1/
1907	M. Planck proposed that mass change in the absorption and emission of heat energy, from the momentum of energy. (He is the first physicist to relate E =mc2 to binding energy,

	estimate molecular binding energy for a mole of water.)
1907	Einstein asserted mass-energy relationship to be true for gravitational energy.
1908	M. Planck proposed that mass-energy relationship holds for all types of energy transfer
1922	Fermi in a paper entitled, "Correction of a serious discrepancy between the theory of electromagnetic mass and the theory of relativity", tried to resolve the difference between electromagnetic mass, $m = (4/3) E/c2$ and $m = E/c2$.

3. From 1925 to 1944 (The proof of E = mc2)

1929	Eddington proposed that the distinction between mass and energy is artificial.
1932	Cockcroft and Walton studied the bombardment of a lithium atom (Li) by a proton (p), which produces two alpha particles (α). **(First experimental verification of E = mc2 and to an accuracy of better than 1%)**
1933	Bainbridge discussed the degree of accuracy of the early quantitative measurements of nuclear reactions which confirmed the mass-energy relationship.
1934	The first photograph showing the creation of a pair of particles, revealed by the fog spots they made in passing through the wet air of a "cloud chamber." The two particles, curving apart under the influence of a magnet, were created in the annihilation of a particle of light. **(First photograph of E = mc2)**

4. From 1945 to 1964 (The shock of E = mc2 and the meaning of E = mc2)

1945	Aug 6: The United States dropped a uranium atomic bomb on Hiroshima, Japan, killing over 100,000. Aug 9: The United States dropped a plutonium atomic bomb on Nagasaki, Japan, killing over 40,000.
1948	Russell pointed out that "atoms" are merely small regions in which there is a great deal of energy. He suggested that "mass is only a form of energy, and there is no reason why matter should not be dissolved into other forms of energy. It is energy, not matter that is fundamental in physics."
1956	The first major nuclear power plant opened in England. **(Application of E = mc2 for daily living)**
1962	Wheeler coined the phrase "mass without mass" to indicate the possibility of removing the term "mass" from the fundamentals of physics.

5. From 1965 to 1984 (The application of E = mc2 in daily lives)

1965	David Bohm suggested that the transformation of "matter" into "energy" is just a change from one form of movement (inward, reflecting to-and-fro) into another form (outward displacement through space). Internal transformations

	taking place on the molecular, atomic, and nuclear levels can change some of this to-and-fro, reflecting "inward" movement, into other forms of energy whose effects are "outwardly" visible on the large scale.
1983	Torretti regarded the terms "mass" and "energy" as designating *properties* of physical systems. The apparent difference between mass and energy is thus an illusion that arises from "the convenient but deceitful act of the mind by which we abstract time and space from nature".
1984	According to Zahar, Einstein showed that "energy" and "mass" *could* be treated as two names for the same basic entity. The apparent difference between mass and energy arises from the contingent fact that our senses perceive mass and energy differently.

6. From 1985 to 1999 (The application of E = mc2 to save lives and the continuous debate on the concept of mass)

1987	Positron Emission Tomography (PET), an imaging system that uses radioactive substances introduced into the body, was introduced. **(Application of E = mc2 in Medical Science)**
1989	Lev Borisovich Okun emphatically declared that in the modern language of relativity there is only one mass, and the concept of relativistic mass is misleading.

1991	T.R. Sandin defended the concept of relativistic mass based on aesthetic reason.
1997	S. Carlip explained that the observational evidence on kinetic energy contributes to gravitational mass, based on general theory of relativity.

7. Beyond 2000 (The future and uncertainty of E = mc2)

2001	Giovanni Amelino-Camelia proposed the third postulate, "Existence of observer– independent skill of mass or length (κ or κ-1) in the Theory of Doubly Special Relativity, and suggested a correction factor for E = mc2: $$E=MC^2$$ $$MC^2$$ $$1+EP$$
2001	Steven Weinberg wrote that "No one knows how to calculate the spectrum of the iron nucleus, or the way the uranium nucleus behaves when fissioning, from quantum chromodynamics. We don't even have an algorithm; even with the biggest computer imaginable and all the computer time you wanted, we would not today know how to do such calculations."
2003	Joao Magueijo claimed that "according to varying speed of light, c is not constant; hence energy or mass cannot be

	conserved." In other words, a varying speed of light allowed for matter to be created and destroyed.
2004	Frank Wilczek in the Nobel lecture, Dec 8, 2004, explained that "Mass comes from energy". (There is energy stored in the motion of the quarks, and energy in the color gluon fields that connect them. This bundling of energy makes the proton's mass.) He described m = E/c2 as Einstein's 2nd Law. Cited in www. http://files.eric.ed.gov/fulltext/EJ848449.pdf

From the above schema on development in $E=mc^2$ by Wong Chee Leong *etal* it is clear that before Einstein, based on the study of the historical development of the concepts, it was possible that the equation $E=mc^2$ would have been discovered by other physicists sooner or later. However, it is likely that Einstein merely accelerated the development (3).

In the second schema, here were at least two main different interpretations of $E=mc^2$. In one interpretation, the relation expresses the inter-convertibility of mass into energy, with one entity being annihilated and the other being created. The other interpretation expresses a proportionality factor between the two attributes, energy and mass, and they were considered to be manifestations of one. In the third schema, Wong Chee Leong *etal* showed how Einstein further developed the philosophical interpretation of $E=mc^2$. There are two realities: matter and field. Field represents energy, matter represents mass. The greatest part of energy is concentrated in matter, but the field surrounding the particle represents energy, though in incomparable smaller quantity. Mass-energy equivalence implies that we can no longer distinguish between "matter" and "the field". According to Wong Chee Leong *etal*, the fourth schema showed that the meaning of E=mc^2 was lively discussed after the landing of atomic bombs on Hiroshima and Nagasaki. The interpretation of "mass can be converted into energy" was suggested to be a misconception. There was also the debate on whether "the law of conservation of mass still holds" and this could be "purely a question of definition." Many authors started writing on E=mc^2, giving the equation its unique and relative meaning.

However, since then, people have been raising questions on the validity of the equation because they still do not fully understand it. More physicists begin to question the validity of E=mc^2. In the modern view, the particle physicists explained that energy is not equivalent to mass. One argument is based on the concept that photon has energy, but no mass.

The Philosophy of the Equation (E=Mc^2)

By Philosophy of E=mc^2, we do not meant that E=MC^2 is a new or latest dimension of Philosophy studied as a discipline in tertiary institutions. But we aim to aptly capture how philosophers of science weight the equation with critical mindset, evaluative power and other logical tools. We shall consider certain critical questions about the equation such as: what are the possibilities of energy changing into mass? Can one observe its accurate state of change? Or can we observe mass moving in a speed of light? How and at what level can these changes occur? Also, do mass and energy mean the same thing? And can it be practically experimented and observed?. From the ethical background, what is the moral justification of the equation?

Ethical Implication of E=MC^2

Ethics is one of the branches of philosophy that studies the moral conducts, theories and principles that exist in a particular society. Ethics investigates the nature, scope and the implications of human code of conducts in the society. Ethics addresses certain moral questions like: what justifies a particular action to be right or wrong? In other words, What are the necessary criteria that justify the right and wrong actions?. Are we justified when we kill our fellow human being(s)? Is there any standard principle of morality? Do we have any reason to do good or bad to our friends? Within the field of ethics, the concept of right and wrong are being evaluated to ensure that human beings live peaceful and integrative. On the question of ethical relevance of the equation from the perspective of its functionality; we posit that, the equation has a negative implication especially as used in1945 to create atomic bomb. By the United States of America, to drop an uranium atomic bomb on Hiroshima, Japan, killing over 100,000 and on Nagasaki, Japan, killing over 40,000. These are really the concern of this paper weighting the equation from its ethical implications. We discovered that Einstein did not personally create the equation, but it was an outcome of his studiousness to uncover the great mysteries of the universe. In other words, through this equation, it was possible for atomic bomb to be produced which resulted in the killing of many human lives in Japan. Ethics here forbids any forms of human conduct that may result in the taking of human lives.

Epistemological Implication of E=MC^2

The greatest task in epistemology is the quest for certainty and truth in knowledge, but due to skepticism it is far from being achieved. Yet theories emerged as an attempt to prepare a bedrock for epistemic justification and essentially to give epistemology a foundation and a means to eliminate skepticism (Mendie,129). The epistemological challenge of this equation is such that even Albert Einstein will not be in a hurry to answer. Such questions may take the shape of provoking our mind on how empirical is the theory in terms of physical relationship between mass and energy. Has one ever observed a mass travelling at the speed of light?

Epistemologically, the equation unraveled the hidden truth between energy and mass. For many years energy was seen to be a separate entity different from mass and vice versa. But since the equation came into science, the property of mass and that of energy were all seen as one and the same thing when moving in the speed of light. This knowledge unlocked the secret of mass and energy, which enable scientists to start a new investigation on the nature, scope and origin of both matter and mass. This brought in a new epistemic understanding in physics and science as a whole.

Logical Implication of E=MC^2

According to Ijiomah, every paradigm of thought or action produces logic and is produced by a logic (*Crisis in Geometry*, 5), meaning every action has its logical interpretation within the context of the logical base. The equation has a logical implication. It is founded on the notion of reasoning faculty, the ability of mass and energy to co-relate and produce energy or mass. It shows a stable logical pattern of good logical relations. Thus, Ijiomah posits that, Logic is the science that seeks to find what should be the correct relationship between or among realities. So that through human reason and by the aid of the knowledge and argument a correct judgment can be reached (Modern, 42). Einstein was able to think using human reasoning power to uncover the natural relationship of mass as energy and energy as mass in a speed of light. What we are trying to say here is that there is a sound logical relation between mass and energy in the equation of E=mc^2. In the absence of this logical relation there could be no product of mass in relation to energy and vice versa.

Metaphysical Implication of E=MC^2

The study of metaphysics is the study of the existence of beings in all ramifications. Being here is seen holistically as whatever exists, what can exist, that is, that which ever has the potentiality of existence is all that metaphysic study. It studies physical beings, spiritual beings, corporeal and incorporeal beings, matter, space, time, momentum; everything one can ever imagine can be insightfully studied under a metaphysical scope. Thus, according to Iroegbu, Metaphysics is:

"That branch of philosophy that studies reality as such i.e. in its most comprehensive scope and fundamental principles. It is the science that tries to determine the real nature of things"(22) $E=mc^2$ is founded on metaphysics or it is metaphysically based because of the composition of matter and energy in the equation. Energy cannot be seen, it is a metaphysical construct that gives matter a meaning in their relationship within the $E=mc^2$ hypothesis. Thus, since metaphysics studies everything, this equation cannot be an exception.

External and Internal Problems of $E=MC^2$

The experimental success of the equation led to the increasing objectification of energy. However, being a matter-motion term, momentum and force, energy neither exists in concrete terms, meaning it cannot be picked up like a piece of stone neither does it move. Glenn Borchardt doubted the existence of energy as such which can be exchanged for mass. He posits that it is simply an idea, a concept, a mathematical description of the motion of matter. Matter does not "contain" energy, for matter only can "contain" other things in motion. Energy is simply a mathematical term necessary for describing and relating the various forms of the motion of matter. He further contends that:

> Many popular accounts maintain that $E = mc2$ describes the conversion of matter into "pure energy," often construed as a kind of matterless motion. Today, "dark energy" and "dark matter" are spoken of as if they were two different "things." Some even hypothesize that the universe was

filled with pure energy before it became filled with matter. This estrangement between matter and motion (separability) is common in popular culture and underlies the 20th century regression from realism to idealism in modern physics. There will be no fundamental change in modern physics until we adhere to the opposing assumption, INSEPARABILITY (Just as there can be no motion without matter, so there can be no matter without motion). Without it, it is impossible to explain the physical meaning of the equation. Like all equations involving aspects of reality, E=mc2 simply refers to the transformation of one kind of matter in motion into another kind of matter in motion and/or the transformation of one kind of the motion of matter into another kind of the motion of matter. (Cited in http://www.scientificphilosophy.com/Downloads/The%20Ph ysical%20Meaning%20of%20E=mc2.pdf)

For him E=mc2 is not possible, as such matter in motion only produces another form of matter-like reality, energy is not a possibility. This same problem cuts across the realist debate on the possibilities of energy being seen as mass. The realist observed that energy cannot be seen as a real entity; rather it is a product of matter in motion. This is why Glenn Borchardt maintains that, like all equations involving aspects of reality, E=mc2 simply refers to the transformation of one kind of matter in motion into another kind of matter in motion and/or the transformation of one kind of the motion of matter into another kind of the motion of matter. Thus, he argued that there is no conversion since matter is a form of energy.

CONCLUSSION

We note here that the foundation of science is based on the growth of idealism. As such the mind has the power to create things that our senses could not have imagined. The equation has its foundation on the ratiocinative power of Einstein's brain and not what he saw using his eyes or other sensory organs. Since when it first came to bear, the equation has not been fully exhausted, its continuous study becomes a necessity. As such, the work has shown the place of philosophy on this famous equation, by insightful exploration of the ethical, logical, metaphysical and the epistemological implications of $E=MC^2$. But it is far from being the end; as such, the equation will continue to flourish within the world of scientific and philosophical evaluations.

CHAPTER SEVEN

QUANTUM MECHANICS AND ITS PHILOSOPHICAL IMPLICATIONS

Introduction

The study of science is geared towards an unending quest for the nature of things around the universe. Science has achieved so much for man right from its inception. The discovery of quantum mechanics in science has become a monumental achievement to all physicist and philosophers in the entire world. Quantum mechanics is a branch of physics that was developed in the first part of 20th century. This theory has achieved a lot of successes, especially in explaining the behavior of atoms, molecules and nuclei. It was developed between 1900 and 1930 and combined with the general and special theories of relativity; it has revolutionized the field of physics. One of the new explanations brought in by quantum mechanics was the explanation of the particle properties of radiation, the wave properties of matter, quantization of physical properties and the idea that one can no longer know exactly where a single particle such as an electron is at any particular instance. This new understanding about the particle universe became necessary in science as it explains all new experimental evidence available in the study of the nature, scope and origin of all subatomic particles.

This paper posits that, the discovery of quantum mechanics was necessary for the development of modern physics. Thus, Planck contribution on blackbody radiation, Einstein's photoelectric effect,

Bohr's 1913 spectra theory, Heisenberg 1927 uncertainty principle were all historic facts about the smooth uncovering of the theory of quantum mechanics which has a lot of philosophical implications. The paper is only an attempt to uncover the scientific and philosophical implications of quantum mechanics which is believed by scholars to be the last theory in science in spite of all other articulations attributed to quantum theory, our focus is limited to the philosophical implications of quantum mechanics.

Conceptual Clarification

In every scholarly work, there is a need for clarification of terms which is of immense importance to readers, to enable an in depth understanding of a research work of this kind. In this work, it is our task to give a brief conceptual explanation to some concepts that are used in this research. A research work is philosophical when there is a greater attention to logical argumentation, critical analysis and proper evaluation. The term "philosophical" from its root explanation is derived from an umbrella word "Philosophy". Philosophy, which is derived from two Greek words *"Philo"* and *"Sophia"* meaning love of wisdom, has enjoyed this etymological definition but, it could also be defined technically as a critical enterprise used in solving the most fundamental problems of human nature. Thus, for a discussion to be philosophical, great attention must be given to the application of some basic philosophical tools like critically, logicality, evaluation etc. to any issue, in attempt to solving problems for that discipline. Quantum mechanics or quantum theory is an area or a branch of mechanics that deals with movement and force in pieces of matter smaller than atom. It is a theory based on the idea that energy exists in units that cannot be divided.

The Prelude to Quantum Mechanics

The journey to quantum theory has not been an easy task during the entire process of its development. To understand this theory, it is imperative to state that one will appreciate the implications of Quantum Mechanics after a good knowledge of Newtonian physics and theory of relativity. Newtonian physics which is also called classical mechanics studies objects at the very large level. In other words, classical mechanics studies macro objects and how these objects behave with other properties of the universe. In classical mechanics, there is a high level of objectiveness and certainty in its explanation on the structure of the universe. Thus, one object can cause the other to behave in a distinct way, and this behavior can be predicted with great certainty as explained by the concept of causality in science. Also, an event can bring forth another event, giving us the impression that all things are interconnected with one another. The idea behind causality states the nature of how event (A) could cause event (B). This process of knowledge makes predictions to be possible. This explains the notion that science is mechanistic and deterministic. In a follow up of this notion Akpan in his work *"Quantum Mechanics and the Question of Determinism in Science"* published in Sophia: and African Journal of philosophy,. Quoted Payel as follows:

> This mechanistic view of the universe brought about a new
> hypothesis of interpreting reality as deterministic.

Underlying determinism is the view that everything in the universe, including all the motion-from the smallest to the largest occur in a way that can be predicted with absolute accuracy using the laws of Newton (Akpan 72).

This view exposes that the structure of science during the Newtonian era was deterministic; his laws of motion were based on the mechanistic view about the universe.

In 1905, Albert Einstein came up with another theory in science to give a revealing explanation to large objects in space and how these objects relate with one another even when in motion. Einstein's relativity gradually gave scientists a better cosmological insight into what quantum mechanics means.

It is imperative to note that Newton capitalized on Galileo's measurement on objects to formulate his laws of motion. For Newton, If a body is not acted upon by any force, it would keep on moving in a straight line at the same speed, or if it were at rest, it would continue to be at rest. The force alters the speed of a body. Speed and the change of speed, the element of direction and inertia all concern the idea of space and time. Newtonian theory of gravity is all that Einstein modified to form his special and general theories of relativity (Sophia, 2007) in 1905. Albert Einstein revolutionized classical physics and turned physics upon its modern course. This revolution lay in the principle of relativity, which guarantees that "all the laws of physics are the same in all inertial reference frames". Einstein took into consideration the features of Newtonian mechanics, the Galiloian transformation, Maxwell's electromagnetic theory of light, the Michelson-Morley experiment and the Lorentz transformation to invent his theory of relativity (Akpan 223). His general relativity arose from the extension of the principle of relativity to the gravitational field (Gribanor, 217). It is the application of relativity to gravity.

In spite of the progress made by classical mechanics and Einstein's theory of relativity there were so many questions that classical mechanics could not account for nor explain. Hence, there was need for another physic. Through Planck's discovery of "quanta" a new science was born which gave us what we now call quantum mechanics.

An Exposition of Quantum Mechanics

Classical mechanics or classical physics was to give way to a new conception of nature and as such, a new physics called quantum mechanics. Gary Zakau (*The Dancing Wuli Master*), defines quantum mechanics as "A quantum is quantity of something, a specific amount. "Mechanics" is the study of motion. Therefore, "Quantum mechanics" is the study of the motion of quantities. Quantum theory says that nature comes in bits and pieces (quanta) and quantum mechanics is the study of this phenomenon" (45). This change about nature in physics occurred at the beginning of the twentieth century. This view was that nature was not simply a blend of forms of matter or physical quantities rather nature was discontinuous and discrete in form of discrete energy called quanta, a remarkable discovery by Max Planck.

Max Plank in 1901 made a remarkable contribution to physics. His was of the view that "energy is not continuous" which he referred to as "Plank constant". The result of this formula was so because energy is always emitted or absorbed in discrete units which he called quanta (Hawking 83). This finding was as a result of his physical observation of the radiation of heated materials which gave him the understating that there is a bit of radiation in atoms, and this radiating atom produces what he called quanta.

His idea of quanta is what modern physicist now calls photon. Thus, Albert Einstein in 1905 interpreted Plank's quantum hypothesis (quanta) to realize certain effect which he used to explain photo-electric effect. Einstein explains that shinning a high dense light on certain material can eject out electron called photo-electrons from such material which can help to form electric current called photo-electricity.

In 1927, a physicist called Werner Heisenberg came out with another theory called "*The Uncertainty Principle*" which shattered the foundation of classical mechanics. It was one of the most important innovations in quantum mechanics, which says that, "the momentum and position of a body cannot be simultaneously measured with unlimited precision" (Wheeler, 14). Broadly speaking, quantum mechanics incorporates four classes of phenomena for which classical physics cannot account to wit: The quantization of certain physical properties, Wave-particle duality, the uncertainty principle, Quantum entanglement, Photoelectric effect, Black body radiation (Mendie 124). This theory shattered the basic assumptions of classical physics by eliminating the possibility of prediction and determinism in science. Thus, quantum mechanics is of the view that, the objective reality as disappeared, reality which brings new theory has evaporated. This is because with quantum mechanics the idea of cause and effect do not exist; as such there is no objective reality which gives formations for empirical theories. Quantum mechanics studies the sub-microscopic realm where the laws of nature do not exist (Mendie 125).

The Philosophical Implications of Quantum Mechanics

In spite of the tremendous successes registered by quantum theory, this paper argues that quantum mechanics has so many implications for science and philosophy which is a positive development to the quest for scientific knowledge, to wit: the introduction of subjectivism to physics, the eradication of the concept of causality, the eradication of determinism in science

The Introduction of Subjectivism to Physics

Subjectivism is a philosophical position that gives reference to personal fancies, feelings, interests and depositions in the acquisition of knowledge. In physics, it was first brought in by Einstein's theory of relativity through his notion of the observer's frame of reference. In quantum mechanics, it is demanded that one is allowed to view the world the way he wants, thus, no objectivity anymore. The objective reality which covers the laws of Newton has evaporated, as such, it is imperative to note that one can create his own unique theories to create a world in which he or she wants to investigate. This is what subjectivism in physics has created in science, this to some extent has led to the emergence of crisis in physics, because of the foundation on which physics was previously built has been faulted by quantum theory.

The Eradication of Causality in Science

Science before the earliest twentieth century had always thrived on causality. Causality can be defined as a philosophical belief that one occurrence can cause another. Hence, in every cause there is an effect. And the effect must have a connective principle to the cause.

The advent of quantum mechanics broke away the core principle of causality in physics. This is because in the sub-atomic realm of quantum mechanics, there is nothing like cause and effect. This is because reality at such realm is so minute, and does not obey the laws of nature where cause and effect take place.

The Eradication of Determinism in Science

Classical science and in fact post-Newtonian science up-till the early twentieth century was driven in a deterministic interpretation of reality. The deterministic hypothesis in science holds that everything in nature has a cause and if one could know the antecedent causes, he could predict the future with certainty (Akpan 270). However, quantum mechanics holds that sub-atomic particles through which the ultimate materials from which all the complexity of the existence in the universe emerges, do not obey deterministic laws, hence, their activities are causally indeterminate and could only be understood with probability.

This view originated from Bohr's view of indeterminacy principle, which states that when an electron in a single orbit, its position cannot be determined. This was further developed by Hersenberg's uncertainty principle about electrons which move from one orbit to another.

The Introduction of Metaphysics to Physics

Philosophy as an all-embracing discipline could be traceable to every aspect of human endeavours. Metaphysics which is one of the core branches of philosophy, and which studies reality in totality, shares some ontological connections with quantum mechanics. During the era of classical mechanics, it was a debatable discussion that anything that cannot be empirically verifiable should be eradicated from science. This was based on the tenets introduced by the positivist movement in modern science.

It is imperative to note that, quantum mechanism investigate the unobservable part of reality which constitutes the sub-microscopic realm of reality to wit: atoms, mesons, photons, spins, electrons. This investigation is only possible through the foundation of metaphysics, because metaphysics studies beings as beings (Beings from its totality). It also follows that since metaphysics is one of the core branches of philosophy that studies both visible and empirical realities, then quantum mechanics can be seen as a science that is studied by metaphysics and by this there is an epistemological link between both of them. Thus, for one to really capture quantum reality, it is not with sense perceptions, this is because our senses cannot penetrate the micro world. It is the position of this paper that metaphysics is beginning to gain more entrance into science which was formally downplayed by the positivist philosophers before quantum physics came into science. What now constitute reality are the unobservable particles like quarks, leptons, spins protons, electrons and plasma, etc. which are all derivable through metaphysical studies of unseen realities.

Conclusion

In summation, it is undisputable fact that since quantum mechanics which studies the unobservable part of reality has a greater link with metaphysical studies; it explains the direct introduction of metaphysical considerations or facts into science. Thus, metaphysics has started claiming back the lost glory that was undermined by the positivist movement in modern science. And as such metaphysics can now be seen as a major tool for all scientific enquiries. This therefore explains further that the ontological essence of African science as founded mostly on non-physical entities has some affinity with the science of quantum mechanics.

The development of quantum mechanics from the Plank's hypothesis of "quanta" to Heisenberg uncertainty and others, who contributed to quantum theory, has help to develop philosophy of physics in connecting subjectivism as a possible implication of quantum mechanics. Subjectivism which introduces relativism to science is capable of bringing forth another reality that will be more useful to physicists, if properly followed.

CHAPTER EIGHT

THE END OF ROAD THESIS IN SCIENCE: A CRITIQUE

INTRODUCTION

On the 15th of December 1920, Neils Bohr and Werner Heisenberg, before the physical society of Copenhagen reveal a distinctive interpretation of quantum mechanics that bewildered the entire scientific community all over the world. That quantum mechanics which studies the sub-microscopic nature of the universe, which is built up with uncertainty and indeterminacy principles is the final theory of physics, hence, *the end of road thesis in science.* In reaction to this assertion, Albert Einstein, a philosopher cum physicist, stood up to raise an argument in 1935, together with Podolsky, and Rosen in their paper entitled, "Can Quantum-Mechanical Description of Physical Reality Be Considered Complete?" Einstein and his colleagues conveyed a message that, what is appearing as indeterminacy and uncertainty, is because we do not know the real nature of reality, so because we don't know it, it appears uncertain and it appears indeterminate, immediately we know the deeper reality, it will reveal that; what is looking as uncertain or indeterminate is because of our lack of knowledge, because "God does not play Dice". We argue that, Einstein's position should be the right position to follow because reality is still unfolding. Thus, to Bohr and Heisenberg, they need epistemic humility to follow the unfolding nature of the universe, because end of science with quantum mechanics could lead us to no growth in science.

In attempt to give an all embracing definition of "who is a man" one may be poised to define man in three conceptual views; man as Homo-sapience (that is man the rational), Homo- amicus (that is man the friendly), Homo-fibre (meaning tool making man). Thus, the Homo-sapience definition of man has gained wider explanation of who a man is, in conjunction with the Homo-fibre conception of a man. It is in these very sense, that man, with all his rational capacity is task to complete his nature through tool making processes. These must be done through the study or knowledge of the physical and natural world base on observation and experiment called science.

Science is aimed at a continuous investigation of the nature of the universe in order to give a rational explanations and further solutions for the benefits of man survival in his environment. These has led science to explore for the ultimate reality in order to account for the basic occurrence in the world so that through observations, explanations could be made in order to further predict the future and life of man in his environment. This has been the foundation of science in the classical-Newtonian age. Where reality was assumed in a concrete nature of explanation, where the activities of science was based on the birth of elementary particles for explanations to be made.

These was the foundation of Democritus notion of the "uncutable" further developed and shaped by John Dalton as an "Atom", which latter took a new dimension to be sub-divided into proton, election and neutron in the works of Rutherford.

However, reflecting the worldview of the Newtonian physics or classical mechanics, one could capture or gain more insight about science, where the science of that age was based on explanation and prediction, hence cause and effect (causality) was at play. Through certain physical observation of the phenomenon, induction is made possible, which further develops or generate to laws and theories. Thus, deductions are established from the laws which end up on explanations and predictions as a mechanism for science of that age. (Chalmers, 8) A.F. Chalmers further represent this view with a schematic explanatory model as dram bellow.

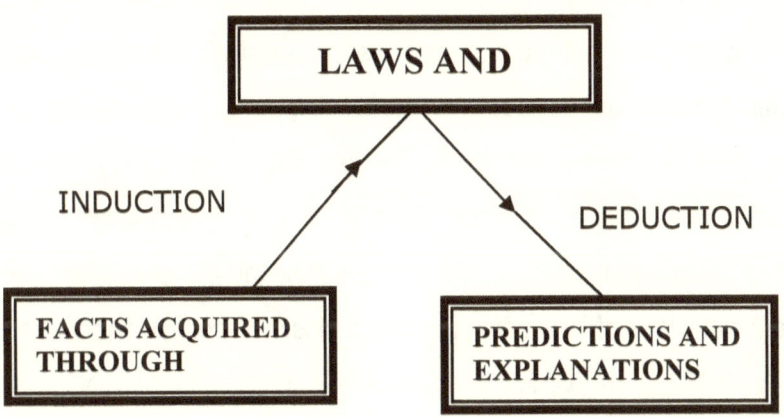

The above schema gave room for Nicholas Copernicus to formulate a comprehensive heliocentric approach of the universe which displaced the Ptolemaic cosmology of the earth as the centre of the universe (geocentricism). Copernicanism had influenced the works of Isaac Newton to formulate his laws

The laws of Newton serve as a base or a foundation for science to grow at the 17th to 19th century cosmology. It was in between this time that the speed of light was discovered in 1876 by Danish astronomer Ole Roemer. Also was the Discovery of the theory of relativity by Albert Einstein in 1905. In 1905 Einstein discovered the special relativity which was based on "frame of reference", that time and space is relative in relation to the observer's frame of reference. In 1916 he published his famous work on General relativity which takes cognizance of gravitation which was not formerly explained in the special relativity. General relativity shows that light is affected by gravity.

However, Einstein Relativity and Newtonian physics had a lot of failures in their time; because they could not give explanations of the sub-atomic realm of reality, the world that exist so minute, the micro realm which are not penetratable using classical mechanics. Any attempt to penetrate the subatomic world using any scientific mechanism effect the true nature of that world. These result on prediction not to be possible in the subatomic world, laws of Newton do not exist at this realm, cause and effect could not be thought of, prediction and determinacy laws of Newtonian physics or classical mechanics could not be possible. But could be explained using "probability" and "indeterminacy" principles, which are guided by quantum mechanics. Classical mechanics or Newtonian physics accounts for explanation basically on activities of the macroscopic or corporeal particles which deal on the inertial and the final stage of a system, without explicit explanation of the intermediate activities of the particles to help the problem of that time. In 1901 Max Karl Ernst Ludwig Planck through his research on quantum behaviour of energy developed a discrete quantity of energy called "quanta". This observation by Planck accounted for the birth of a new physics called quantum mechanics.

Oxford Dictionary of chemistry defines Quantum mechanics as:

> "That branch of physics which deals with the behaviour of matter at the level of the atom, the nucleus, and the elementary particles. At this level, energy, mass, momentum and other quantities do not vary continuously, as they do in the large-scale world, but come in discrete unites called quanta"

It can further be seen as a mathematical theory, as a replacement for classical mechanics in order to explain satisfactorily the behaviour of atoms, molecules, and elementary particles in terms of observable quantities such as the intensities and frequencies of spectral lines. Quantum mechanics changes or revolutionized the understanding of nature at that age up till this present time. Every discipline adjusted their syllabus in-line with quantum mechanics this is why we, have, quantum chemistry, quantum electronics, quantum optics, quantum logic, quantum information science etc.

On the account that, Quantum mechanics gave a clear explanation of what Newtonian or classical physics could not, Neils Bohr and Werner Heisenberg the chief contributors and foundationalist of quantum mechanics on the Copenhagen meeting avers that, quantum mechanics is "the end-of-road-thesis in science", the final theory, the never-to-be surpassed revolution in physics (Egbai, 14). This implies to

Bohr and Heisenberg that no other theory will supersede quantum mechanics, in fact "science has come to an end". This is because they avers that, the objective reality has evaporated, reality which brings new theory has evaporated because, when you look out of the outside world, you look at the things as they are, you begin to gather fact, test hypothesis, formulate theories, then one can think of new theory, that kind of world exist no more, the reality at the level of sub-atomic particles does not obey laws of nature, were we have the laws of cause and effect. In the sub-microscopic world of quantum mechanics there is nothing like cause and effect, but we can only predict in probability and not with certainty. This gave rise to the theory of indeterminacy and uncertainty.

Einstein with other scientists like Podolsky, Rosen etc, never agreed the position Bohr and Heisenberg claimed about quantum mechanics. Though, Einstein had contributed in the development of quantum mechanics on "photoelectric effect". In reaction to this, Einstein spent a large portion of his career arguing against quantum mechanics as the last theory of science, despite his contribution on quantum theory development.

This was viewed on his work in 1935 published together with Poldolsky, Einstein and Rosen (EPR) called *"Can Quantum-Mechanical Description Of Physical Reality Be Considered Compete?"*. Einstein never agreed with Bohr's view, he avers that quantum mechanics is not complete", that a deeper search of reality will emerge. Another theory greater and more encompassing that will surpass quantum mechanics will be born. Thus Einstein keeps telling Bohr and Heisenberg that "God does not play dice". This view by Einstein did not agreed with the Copenhagen interpretation of quantum mechanics. The Copenhagen interpretation supported the completeness of quantum mechanics in science because quantum mechanics gave explanations of what classical mechanics could not which we will capture in a latter section of this work.

Thus, the attitude of Neils Bohr and Werner Heisenberg who where the key advocators of the end of science with quantum mechanics constitute a major problem in science which led to a disagreement with Albert Einstein, Podolsky and Rosen to argue on what should constitute the end of science and not quantum mechanics. Two camps emerged; one maintained that quantum mechanics could explain clearly what Newtonian or classical mechanics could not, on these regard, quantum mechanics is the final theory, the never-to-be surpassed revolution in science.

The other camp led by Albert Einstein maintained that what has been paraded by Bohr and Heisenberg as the end of road thesis in science, does not deserve any of such name at all, and that science cannot come to an end with quantum mechanics because a deeper search of realty will emerge another theory greater and more encompassing that will surpassed quantum machines. Quantum mechanics which support probability, according to Einstein any theory which considers probability should not be an end theory, because for him "God does not play dice".

This debate presently seems to have been laid to rest because of the Copenhagen interpretation of quantum mechanics, where their focus was that quantum mechanics is a complete and last theory of science. Many disciplines have married quantum mechanics as the foundation of their enquiring, which they have parted into this great view of Bohr and Heisenberg.

In this paper we try to address the question, Is quantum mechanics the final theory in science, as admitted by Bohr and Heisenberg?. The attitude of Einstein's ideas and focus for another theory which will surpass quantum theory also constitute the problem upon which this paper focuses.

However, this work is a "rejoinder" or a "retort" of the fundamental fact claimed by Einstein. We accept and deposit that Einstein and the E.P.R. counterparts were right. Despite the fact that, quantum mechanics have improve the world through rapid development of modern technology etc. but it does not support holism, hence could not be describe as a complete theory. We further establish the need for epistemic humility, which will provides us an enabling scenario to understand reality, because knowledge about reality is a continuum. Quantum mechanics as the last theory of science as deposited by Bohr and Heisenberg will mark an end to the whole epistemic enquiry about nature which will cause the search for more knowledge to come an end. Through epistemic humility in quantum mechanics we advocate that there is a deeper theory out there that will surpass Quantum mechanics.

AN EXPOSITION OF QUANTUM MECHANICS

Asking a question, why do we need quantum mechanics? This will agitate our conscious mind to articulate that, there was a problem with classical mechanics, hence the advent of theory of relativity through the Copernican–Galilean revolution could nether explain. Despite the fact that classical (Newtonian) mechanics works perfectly in explaining the world around us, and is accurate amount and even charting the trajectory of probes sent to Jupiter and beyond (Wikipedia.com). The question is why are we not contented with classical physical? Where does the need for quantum theory arise?

However, it was observed that up till the end of the nineteenth century matter was believed to have a continuous existence. This is why the physicists of this era conceived that the natural world is a continuum, and simply in a blend of forms of matter. This created a lot of problems for the classical physicist. The classical theory is unable to explain experimental observations regarding small object which exist in the sub-atomic realm of reality. They could not give explanations concerning the activities of the sub-atomic world. Others include:

i.   Discrete spectra emitted by excited atoms.

ii. Photoelectric effect

iii. Variation of heat capacity of mono-atomic solid with temperature

iv. Spectral distribution of energy in black body radiation etc.
     (Shashi Q C.1).

In attempt to solve the above problems, classical physics was to give way to this new conception of nature and so, a new physics called quantum mechanics was born. According to Gary Zukau (The Dancing Wuli master), he defines quantum mechanics as:

A "quantum" is a quantity of something, a specific amount. "Mechanics" is the study of motion. Therefore, "Quantum mechanics" is the study of the motion of quantities. Quantum theory says that nature comes in bit and pieces (quanta) and quantum mechanics is the study of this phenomenon (45).

This radical change from the longstanding view about nature occurred in the beginning of the twentieth century. The radical view was that nature was not simply a blend of forms of matter or physical quantities into a smooth, continues existence. Rather, nature was discontinuous and discrete. It is a new dimension of science that evolved during the first new decades of the 20th century in an endeavour to understand the fundamental properties of which matter is made up of. It started with the study of the interaction of radiation and matter, with the use of classical mechanics, radiation effect could neither be explained, nor the theory of electromagnetism. In particular, physicists were puzzled by the nature of light peculiar lines in the spectrum of sunlight had been discovered earlier by Joseph von Fraunhofer (1787-1826). These spectral lines were systematically catalogued for various substances, yet nobody could explain why the spectral lines are there and why they could differ for each substance. It took about one hundred years, until a plausible explanation was supplied by quantum theory.

In a contrast examination to Einstein's relativity, which is about the largest things in the universe, quantum mechanics or quantum theory deals with the tiniest thing we know, the particle that atoms are made of which is call "subatomic particles". It imperative to note that quantum mechanics was not the work of one individual, but the collaborative effort of some of the most brilliant physicists of the 20th century, among them are, Niels Bohr, Erwin Schrödinger, Albert Einstein, Wolfang Pauli, Max Born, Max Planck (1858-1947) and Werner Heisenberg (1901-1976) it has been on record that, Planck is the originator of the quantum theory, while Heisenberg formulated one of the most eminent laws of quantum mechanics, which is the "uncertainty principle" which is occasionally also referred to as the "Principle of indeterminacy".

Max Planck remarkable contribution was that "Energy is not continuous" which is referred to "Planck constant". It all happened around 1900 at the university of Kiel were he concerned himself with observation to the radiation of heated materials which he attempted to draw the conclusion from the radiation to the radiating atom on the basis of empirical data, he developed a new formula which later showed remarkable agreement with accurate measurement of the spectrum to heat radiation. The result of this formula was so that energy is always emitted or absorbed in discrete units, which he called quanta (Hawking 83). He further developed his quantum theory to derive a universal constant, which came to be known as "Planck constant". The resulting law states that the energy of each quantum is equal to the frequency of the radiation multiplied by the universal constant $E=hv$. The discovery of quanta revolutionized physics, because it contradicted conventional ideas about the nature of radiation and energy. Planck's hypothesis that energy is radiated and absorbed in discrete "quanta" (or energy element) precisely matched the observed patterns of blackbody radiation. According to Plank each energy element E is proportional to its frequency v: E =hv (Ayi, 405) where h is Planck constant. Thus, the 1905 Albert Einstein interpretation of Planck's quantum hypothesis realistically was used to explain the photoelectric effect in which shining light on certain material can eject electron from the material. Albert Einstein serious contribution further development shows that an electromagnetic wave such as light could be described as particle (later called the photon) in 1926 through the work of Bohr and Heisenberg. They published results that close the "Old Quantum theory", which explain that discrete quantum energy of photon is independently on its frequency. According to Egbia; modern inventions such as television camera tube, solar power cell and photographic light meters all relate to Einstein's description of the photoelectric effect (14). Hence further development of unity theory between the atomic particle and electromagnetic waves called wave-particle duality, in which particles and waves were neither one nor the other, but had certain properties of both was accomplished. Quantum mechanics described the world of the very small; it explains and investigates macroscopic system such as superconductors and super fluids which classical mechanics could not explain. Quantum mechanics through it modus Operandi and experiment, repeatedly produce result which the physics of Newton could neither predict nor explain; yet although Newton's physics could not account for phenomena in the microscopic realm. Quantum mechanics is based upon experiment conducted in the subatomic realm. As Zukau may asset, it predicts probabilities, these probabilities phenomena subatomic phenomena. Subatomic phenomena cannot be observed directly, none of our senses can detect them, not even any scientific apparatus can tell us exactly its true state, because the subatomic world keeps changing and the apparatus itself keeps defecting it true nature. It is pertinent to note that not only has no one ever seen an atom much less no one ever seen an atom, much less an electron, no one has ever tasted, touched, heard, or smelled on ether. This is what quantum mechanics can explain. It is the new dimension of science; Quantum mechanics claims that the deterministic view of reality does not hold in the quantum world. The atom seen as the building blocks by the classical physicists has been split into further minute particles like quarks, spin, photon and leptons etc, which is said to be more than a million times smaller than the head of an office pin (Akpan 72). Thus, neither the movements nor the position of the subatomic particles can be measured or predicted with certainty. However, it is observed that what holds at this level is randomness and chaos, which can only be explained by the "uncertainty principle" or the "indeterminacy principle", hence only probabilities are precisely determined in quantum world. This makes quantum theory, or quantum mechanics to shatter the three basic assumptions of classical science which includes:

i.   Causality
ii. Determinism
iii. Objectivism.

It is pertinent to note that atomic particles are the ultimate materials from which all the complexity of existence in the universe emerges, do not obey deterministic and prediction laws, hence their activities are causally indeterminate and could only be understood with probability.

However, classical mechanics works very well at this present time for large objects that are moving slower than the speed or velocity of light, on the other hand when object are so small quantum theory is needed for a better explanation. This is why in quantum mechanics, some of these behaviours are macroscopic and only emerge at extreme low or very high temperatures or energies. This is why Quantum mechanics is derived from the observation that some physical quantities can only change in discrete amount called "quanta" and not in a continuous analog way. For example the angular momentum of an electron bound to an atom is quantized. In the content of Quantum mechanics, the wave-particle quality of energy and matter and the uncertainty principle provide a unified view of the behaviour of photons, electrons, and other atomic-scale objects as formulated by Rutherford which classical physics could not. Many of the results of quantum mechanics are not easily visualized in terms of classical mechanics. For instance, the "ground state" in a quantum mechanical model is a non-zero energy state that is the lowest permitted energy state of a system, as opposed a more "traditional" system that is thought of as simply being at rest, with zero kinetic energy instead of a traditional static, unchanging zero state, quantum mechanics allows for far more dynamic chaotic possibilities, according to John Wheeler in spooky quantum.

Broadly speaking, quantum mechanics incorporates four classes of phenomena for which classical physics cannot account:

i. The quantization of certain physical properties

ii. Wave-particle duality

iii. The uncertainty principle

iv. Quantum entanglement

v. Photoelectric effect

vi. Black body radiation. (Wikipedia.com)

Schematically, the following are the important contributors of the foundation of quantum mechanics and principles they uncovered.

YEAR	RESEARCHER	CONTRIBUTION
1901	MAX PLANCK	Blackbody Radiation
1905	ALBERT EINSTEIN	Photoelectric effect
1913	NEILS BOHR	Spectra theory
1922	COMPTON	Photo scattering
1924	WOLFANG PAUL	Exclusion principle
1925	De BROGLIE	Matter wave
1926	ERWIN SCHROEDINGER	Wave equation
1927	WERNER HEISENBERG	Uncertainty Principle
1927	DAVISON & GERMER	Wave properties electrons

WERNER HEISENBERG'S AND NEILS BOHR'S ARGUMENT ON QUANTUM   MECHANICS AS THE END OF ROAD THESIS IN SCIENCE

The argument raised by Heisenberg and Bohr was a debate in which they engaged, was surely one of the monumental debates of the 20th century. There exist two groups of titans of modern physics with quite opposed position, struggling to establish their view of the meaning of quantum mechanics and it completeness in explaining the universe.

Niels Bohr one of the major contributors of quantum mechanics, has being a very difficult task to achieve its underlying fact, and thus establishing it took a lot of physical and intellectual combat to reached its ultimate goal. He never got rid of the difficulties encountered. All his success most time landed in a very confusion of cognition, lack of comprehensive ideas, and lack of clarity. Bohr implies to himself "oh great knowledge of the world of science one must reach with great success". Bohr was task to arrive at what will give a better explanation to the scientific problems of his time. Bohr later got out of this confusion to gain a full knowledge of quantum mechanics as the "all and all thesis in science", the end, and the final road of which science can never reach. Heisenberg further explain, that quantum theory came to existence because it was clear to all scientist that at the atomic level one can easily understand the particles, waves movement, but at the subatomic realm reality was no more visible. Thus, there was now limit to which our understanding can attain; prediction, explanation and observation was now based on probability. Reality in which science is investigating is no more out there, we are now the maker, the creator of our world, objective reality has disappeared, it is the individual that now create his world in which he want to investigate on using quantum mechanics. Bohr and Heisenberg challenges the fact that classical mechanics claim to give explanations of the macro world can never be certain of how valid those explanation could be. This is because, "it is not the way we see the world that the world really is". Bohr further explained that, what one observer derives from observation defers from what others may observe; this is because at the quantum level there is no certainty neither prediction that is possible.

These led Bohr immediately to what seems to be the central idea; as he avers:

> This crucial point, which was to become the main theme of the discussions, is reported in the following, implies the impossibility of any sharp separation between the behaviour of atomic object and the interaction with the measuring instruments which serve to define the conditions under which the phenomena appear. Infact, the individuality of the typical quantum effect finds it proper expression in the circumstance that any attempt of subdividing the phenomena will demand a change in the experimental arrangement introducing new possibilities of interaction between object and measuring instruments which in principle cannot be controlled (*Standard Encyclopedia of Philosophy*)

Bohr explains from the above that there exist an interaction between the micro substance, instrument, and the observer which do not give room for objectivity, or making prediction to become possible. Hence, can only be explains using the "indeterminacy principle" of quantum mechanics. This is because reality there is so minute and a mere attempt to observe them changes their true nature from what they are, hence cannot be predicted or determined. Bohr advocated for a complementarity within the interactive world of experimental procedures. Since the individuals are the creator of his own world, there was need for all scientists to bring forth all their result into one interactive section to foster unity in science, Bohr calls this the "notion of complementarity" Bohr notion of complementarity is to bring together the probabilities of various outcomes this explains that there is no objectives physical reality other than that which is revealed through measurement interpreted.

However, Quantum mechanics had enormous success in explaining many of the features of our world. The individual behaviour of the subatomic particles that make up all forms of matter (electrons, quarks, protons, neutrons, photons, spins, leptons etc) can often only be satisfactorily described using quantum mechanics; hence the interpretation of scientist that quantum mechanics the Copenhagen interpretation.

EINSTEIN, PODOLSKY AND ROSEN'S ARGUMENT AGAINST QUANTUM MECHANICS AS THE END OF ROAD THESIS   IN SCIENCE

The Copenhagen interpretation of quantum mechanics as the end theory of science by Niels Bohr and Werner Heisenberg did not go well with Einstein. Einstein who was the major contributors to the development of quantum mechanics never accept this view, he leaved his entire career auguring against Bohr's and Heisenberg's interpretation of quantum mechanics. He argued that probability in quantum mechanics reflect lack of knowledge in science, and not the end of science. According to Einstein he avers that; the world cannot be ruled by "randomness" and "chaos", because for him "God does not play dice". He advocated for orderliness in the nature of the universe and not with a theory that facilitate only "uncertainty principle" or the "indeterminacy principle", hence on a deeper investigation into nature may emerge another theory more encompasses than quantum mechanics.

In the works of John D. Norton on *"Einstein on the completeness of quantum theory"*, he deposited that Einstein had recognized that this new quantum theory enjoyed remarkable empirical successes. However, he did not believe that future fundamental physics should be build upon it. Rather he thought the way ahead was to develop the geometrical approach of general relativity into an all encompassing "unified field theory" within which the results of the new quantum theory would he derived. Because of the failures of Einstein in attaining the marriage between general relativity with quantum theory, he became the major critics and the most prominent physicist to criticize the new quantum theory. According to Karl popper in his book *Quantum Theory and the Schism in Physics*, he avers that

> Bohr too was, of course, a passionate admirer of special relativity theory. He would have wanted to avoid rejecting it, like almost everybody in those days. Had it been shown that such rejection would be necessary if we wish to uphold

quantum mechanics, it may well have meant, even for Bohr, the rejection of quantum mechanics. For special relativity more or less set the standard to which quantum mechanics had to conform. (30).

In all this, Einstein was defending a minority view in the physics community. The task of responding to Einstein was taken up by Bohr. The debate in which they engaged was surely one of the monumental debates of the 20th century. Here were two opposite opposed positions, fighting to establish their assertions of the meaning of quantum theory.

THE EPR DEBATES AGAINST QUANTUM THEORY

In 1935, Einstein, Podolsky and Rosen (EPR), reacted against quantum mechanics, they were of the view that reality can never be explain at once and finally by a single theory, they further avers that nature is still in the process of evolution and that quantum mechanics is only a process of getting an understanding of a better reality, and never the end, the final theory of nature. This was published in one of their famous paper called "*Can quantum mechanical Description of Reality be considered complete?*". In this paper Einstein struggled to the end of his life for a theory that could better comply with his idea of causality, protesting against the view that "there exists no objective physical reality" other than that which is revealed through measurement interpreted in terms of quantum mechanical formalism. However, Albert Einstein and his colleagues Boris Podolsky and Nathan Rosen which are collectively known as EPR, designed thought experiment intended to reveal what they believed to be inadequacies of quantum mechanics.

The EPR position was against the probabilistic understanding of nature, using quantum mechanics. According to quantum mechanics, under some condition a pair of quantum system many are described by a single wave function, which encodes the probabilities of the outcomes of experiments that may be performed on the two systems, whether jointly or individually. At the time the EPR article was written, it was known from experiment that the outcome of an experiment sometimes cannot be predicted uniquely. The EPR paper written in 1935 further shows that the explanation of Heisenberg's uncertainty principle is inadequate. According to standard Encyclopedia of philosophy considering two entangled particles, let's call them A and B, and Pointed out that measuring a quantity of a particle A will cause the conjugated quantity of particle B to become undermined, even if there was no contact, no classical disturbance thus, in Heisenberg's principle, it was an attempt to provide a classical explanation of a quantum effect sometimes called non- locality. According to EPR there were two possible explanations. Either there was some interaction between the particles, even though they were separated, or the information about the outcome of all possible measurements was already present in both particles. The E.P.R avers in summation that since the distance and momentum of a particle in an orbit cannot be measured simultaneously, but it is clear and certain that those features were already present in them. This is why their view is always regarded as "The E.P.R. paradox" (wikipedia.com). Since there was such limit in quantum mechanics about the understanding that particles "Distance" and "momentum" do exist certain values, if measured simultaneously that will give us certainty in figures.

They then concluded that quantum mechanics was incomplete since in its formalism, there was no space for such hidden parameters.

In summation, according to John Norton, Einstein never accept quantum theory throughout his life time on earth, despite the fact that he embrace quantum mechanics as a new dimension of physic but not as the end theory of science. Einstein referred to this situation in his oft repeated quip that he could not believe that God play dice. The remark seems to have been made frequently, but mostly in conversation. Here is how he put it:

> Quantum mechanics is very worthy of regard. But an inner voice tells me that this not yet the right track. The theory yields much, but it hardly brings us closer to the old one's secret. I in any case, am convinced that he does not play dice..., it is hard to sneak a look at God's cards. But that he would choose to play dice with the world ... is something I cannot believe for a single moment of my life (Einstein, 234)

Thus, probability in quantum mechanics is just like a dice, hence the world ruled with dice. Einstein pursued this project for decades, up to his death. However, the final results were inconclusive. As he dug himself deeper into these investigations, the mainstream of physics turned in other directions. While Einstein was struggling to understand how to unify two forces, gravity and electromagnetism, physics had discovered two more fundamental forces, the weak and strong nuclear forces. And while Einstein focused on the geometrical approach that proved fruitful in the 1910s, quantum physicists were dealing with a new theory in which the idea of an observer independent reality was becoming elusive.

However, since Einstein's death, experiment analogous to the one described in the EPR paper have been carried out, starting in 1976 by French scientist Lamehi – Rachti and Mitting at the Saclay Nuclear Research Centre. These experiments appear to show that the local realism idea is false, thereby supporting the position of Bohr etal., against the challenge from Einstein and his group. Einstein died with his assertion poor days of science for it is our ignorance of these smoothed away properties that makes a probabilistic assertion the best we can know. Einstein has a ready explanation of the probabilities that have now entered physics in quantum measurement processes; they are merely expressions of our ignorance.

A PHILOSOPHICAL DEFENCE OF EINSTEIN'S' VIEW ON QUANTUM MECHANICS AS THE END OF SCIENCE

Albert Einstein a philosopher and a physicist had contributed numerously in science which help on the speedy development of quantum mechanics, was been over thrown by the subsequence position of Bohr and Heisenberg. Bohr and Heisenberg took quantum mechanics to a new dimension that disrupts the foundation of which classical mechanics was built. Famous of this development was the uncertainty and indeterminacy hypotheses that one event can cause another (causality). Einstein never gave up to the fact that quantum mechanics is the last theory of science, the end of road thesis in science. But he raised a position that, the world could not be rules with randomness and chaos hence the need for orderliness in science. He spend all his career arguing against Bohr and Heisenberg's position on quantum theory, this is because for him "God can never play dice".

In defence of Einstein position, we strongly think that Einstein's view on the statue of quantum mechanics should be the right deposition while investigating nature. In line with this argument, it is noted that Einstein did not think there could be any theory that could be the last theory in science because reality is still unfolding time to time, thus in support of Einstein,Uti Egbai avers that:

> It is first of all important to remember that Einstein never regarded any of his own theory as a final breakthrough. His own photon theory and the need to use it together with the wave theory of light, which really established what was later called wave-particle duality, he regarded as a stop gap, although it brought him almost to the threshold of the theory of matter wave;... His special theory of relativity he also regarded, rightly as unsatisfactory for several reasons, empirically because it merely replaced absolute space by the absolute set of inertial systems. Of his general theory of relativity, he said that "ephemeral" and from the moment of its conception to the end of his life he tried to supersede it (17).

Thus, there is all possibility that theories are raised, which with time do fall. This is because reality can never be conceived at once because of human limitation. Hence, indeterminacy and uncertainty in nature is because of human limitation to attain the acquisition of full knowledge about nature. When once a theory is established, there must be a time that another will emerge and probably will have a superior content increase than the former; this was the line of Einstein thinking.

However, Einstein was right to most have raised a position against Bohr and Heisenberg, because in the science of today, there is a continuous development from one theory to another. On 4th of July 2012, a particle called Higgs boson-like particle was discovered. The particle has been the subject of a 45-year hunt to explain how matter attains its mass. This was announced at the European organization of Nuclear Research (EONR) in Geneva. There is need to accept that knowledge is a continuum. Hence any attempt to end all theory with quantum mechanics will lead scientist epistemic enquiry to any end. This will hurt human ability to seek for more explanation, more investigation of nature. Science is an investigatory discipline; the Higgs Boson-like particle discovery could have not being possible if all scientists relents all their effort on the theory of quantum mechanics.

It is interesting to note that classical physics still has a lot of role to play in our life. Scientist still enjoys the hypothesis of cause and effect. Macro world still play salient role in human cognitive ability. This is why there is a lot and numerous inventions in science. Science, more than anything else, was Einstein's life; and to understand the man, it is necessary to follow his scientific ways of thinking. He never at any time rejects the impact and implication of quantum mechanics, quantum mechanics to a large extent gave explanation of what classical mechanics was not able to explain. Explanation of such includes as follows:

1. Black-body radiation
2. Photoelectric effect
3. Wave-particle duality etc

It is imperative to follow by careful investigation of the impact of Einstein that effect the speedy discovery and development of quantum mechanics. In 1905 Einstein published five scientific papers, including the special theory of relativity and an addendum that said that the energy content of a body is equal to its mass times the velocity of light squared ($E=MC^2$). This discovery for Einstein was never the last theory in physic. He became more fateful to the investigation of nature. Still that same year (1905) he made a mew discovery in science on photoelectric effect that gained him the 1921 Nobel Prize in physics. Einstein still gathered more enthusiasm and zeal that propelled the discovery of the general theory of relativity in 1916.

All these discoveries for Einstein were never the final or end of road in science. This is why with the advent of quantum mechanics; it was Einstein thought that it could both be married with his theory of relativity to attain what is called a "unified field theory". Bohr and Heisenberg, quantum mechanics has explain everything about the nature and scope of the universe, all other theories are subjected to quantum mechanics including Einstein relativity. This is why they deposited that quantum mechanics is the end of road thesis in science, perhaps the last theory of physics. Looking at Einstein in retrospect, one can picture the fact that, he was so zealous for a continuous search of a theory that is more encompassing than quantum mechanics, that is why the marriage between quantum mechanics and relativity was his major goal.

This had gave the scientific community a long lasting debate of all time, but won by Bohr, because Einstein attempted to produce a unified field theory failed. Einstein failed not because quantum mechanics is the last theory in physics, but because his unified field theory never march the orbit of quantum cosmology.

The quest for a "unified flied theory" which could give room for the establishment of an underlying link between the seemingly unrelated force of gravity and electromagnetism initiated by Einstein was to solve the problem of the vision of marring general theory of relativity with sub-microscopic quantum world of quantum mechanics failed. He admitted in his book *Out of my Later Years* that: "I must seem like an ostrich who forever buries its head in the relativistic sand in order not to face the evil quanta" (13).

Thus, in further support of Einstein position, we render a retort to Bohr and Heisenberg that reality is still unfolding and that when we generate full knowledge about nature thereby gaining deeper knowledge, there is all possibility that a new theory will soon emerge. This is why the Giggs-Boson- like particle is discovered which helps, to give explanation of how matter attains its mass. This means that reality is still unfolding and knowledge is a continuum. The new generation of physicists has it that reality cannot be attained with particle-like or quark's-like but with infinitesimal loops that resemble tiny vibrating strings. This framework is popularly known as string theory. This is to the effect that nature or reality is still unfolding when deeper knowledge of nature is gotten as projected by Albert Einstein. He is the cause of the major growth of science, which he gave room for more investigation. In 1995 it was discovered that string theory outside the four dimension (height, time, width, and length) consist also a new class of subatomic particles known as "supers-metric particles" or "sparticles". Hence, it represent a deeper reality still unfolding which scientist call M- theory, M- theory stands for many things including matrix mystery, magic and murky, which means. Einstein was right to argue that quantum mechanics is not the end of road thesis as projected by Bohr and Heisenberg.

If Einstein was alive, there is no possibility that he could accept m-theory as the last theory in physics, because his life was an investigative one, where there is a deeper knowledge for man, the result will be a new unfolding reality that will bring another surpassing theory, meaning that knowledge is a continuum and a journey of no end.

SUMMARY AND CONCLUSION

Science, until the beginning of the twentieth century had mostly followed by the train of observation, gathering of fact, forming of hypothesis, formation of theories and formations of law, which gave room for determinacy and certainty in science. Where, it was belief that one event can cause another, in line with cause and effect principle. The advent of quantum mechanics brought a change to this line of reasoning. Bohr and Heisenberg theories of indeterminacy and uncertainty broke the foundation in which classical science was built upon; hence quantum mechanics as the end theory of science, the never to be surpasses theory of physics. Bohr and Heisenberg proposed that, the objective reality has evaporated, that reality which brings new theory has evaporated because when you look out of the outside world, you look at the things as they are, you begin to gather fact, test hypothesis, formulate theories, then one can think of new theory, that kind of world exist no more, the reality at the level of sub-atomic particle does not obey laws of nature, were we have the laws of cause and effect. In the sub-microscopic world of quantum mechanics there is nothing like cause and effect. We can only predict in probability and not with certainty. This gave rise to the theory of indeterminacy and uncertainty in quantum mechanics. Einstein never accepts the position of Bohr and Heisenberg, he avers that; what is appearing as indeterminacy, what is appearing as uncertainty, is because we do not know it, it appears uncertain and it appears indeterminate. And immediately we know the deeper reality, the indeterminacy and the uncertainty will be reveal that, what is looking as uncertain or indeterminate is because of our lack of knowledge. Einstein is saying that "God cannot play dice". Heisenberg and Bohr keep remembering Einstein "don't tell God what to do, because God can choose to play dice". In sympathy with Einstein's position we argue that reality is still unfolding, that quantum mechanics is not the last theory of science.

We argue that Einstein position was the right deposition while investigating nature, because knowledge is a continuum. Though he failed on the debate with Bohr and Heisenberg, but we consider the fact that nature is dynamic and not static. This is why man by nature is full of energy and new ideas, once we gain a deeper understanding about reality; it implies the creation of a new theory that will surpass quantum mechanics. This can be attested by the discovery of string theory that further generate to m- theory and the present discovery of the Higgs boson-like particle on 4th of July 2012 at Geneva.

One can never doubt the fact that quantum mechanics has lifted science to a stage which was never thought of with classical mechanics. It is imperative to note that Bohr and Heisenberg's position of quantum mechanics if followed could lead science to stagnation. Whereas, Einstein deposition about quantum mechanics, if followed, is more capable of leading science to a new knowledge since it encourages continuous research.

However, we argue that Neils Bohr and Werner Heisenberg were wrong because reality is still unfolding. In other worlds, Bohr and Heisenberg needed epistemic humility in studying the nature of the universe. They must be ready to humble themselves when studying the unfolding nature of the universe, so as to follow changes of nature steps by steps, and not only with quantum mechanics as the final theory of science.

This is why today, scientist are dealing with string theory, m-theory and others yet to be observed because Einstein was right, in the sense that one theories that are uncertainty cannot rule the world because God does not play dice.

CHAPTER NINE

STRING THEORY: A REALISM OR IDEALISM

Introduction

String theory currently is the only theory vying for the total explanation of all known entities, the world of matter, theory of relativity and quantum mechanics. This theory predicts that all forces of the universe could be unified into a single manipulation of tiny loops of vibrating energy called strings. The long quest for this theory has been a demanding development to search for the theory of everything by modern physicists which started with the works of Albert Einstein to unify his theory of relativity with quantum mechanics called quantum gravity that made it possible to view reality as one dimensional string. This paper presents the various predictions in the development of string theory, the discoveries that could merge every reality together. We shall further articulate the various philosophical implications of string theory from the realist and the idealist perspectives as it concerns the arguments of whether it describes an existing reality or whether it is an idealistic construct with no concrete and/or tangible expression in nature. But in this paper, we articulate that, the assumptions supported by idealism suit how science grows, because the mind is central to human understanding of the unfolding nature of the universe.

If one is asked to explain what is the composition of the universe or what makes up the universe, most times, one will be tempted to make a list of the stars, moon, sun, oceans, rivers, mountains, buildings, air, and momentum and so on. These are things that can be penetrated within the ambience of our senses. It was this direction of observational medium that influenced the discoveries of Galileo, Isaac Newton and so many others to perceive reality as only a product of the five senses, the sense of sight, sound, touch, smell, and taste. These were factors that formed the building blocks of matter within the science of Newtonian science. Science for Newton is empirically based and it is a continuous investigation of the nature of the universe in order to give a rational explanations and further solutions for the benefits of man's survival in his environment. These have led science to proceed in the spirit of exploration for the ultimate reality in order to account for the basic occurrence in the world so that through observations, explanations could be made in order to further predict the future and life of man in his environment. This has been the foundation of science in the classical-Newtonian age. In the world view of Newton, reality was assumed to have a concrete nature with physicalistic explanation. Here the activities of science were based on the birth of elementary particles for explanations to be attained. It was this way of perceiving reality that led Democritus to drive to search for the notion of the "uncutable" which was further developed and shaped by John Dalton as an "Atom", which latter took a new dimension to be sub-divided into protons, electrons and neutrons in the works of Rutherford (Mendie 120).

However, reflecting the worldview of the Newtonian physics or classical mechanics, one could capture or gain more insight about science. Science of that age was based mostly on empirical explanation and prediction, hence cause and effect (causality) was the predominant mode of judgment. Through certain physical observation of the phenomenon at the macro level, induction is made possible, which further develops or leads to laws and theories. Thus, deductions are established from the laws which end up as explanations and predictions as a mechanism for science of that age. (Chalmers, 1990: 8) A.F. Chalmers further represent this view with a schematic explanatory model as shown bellow.

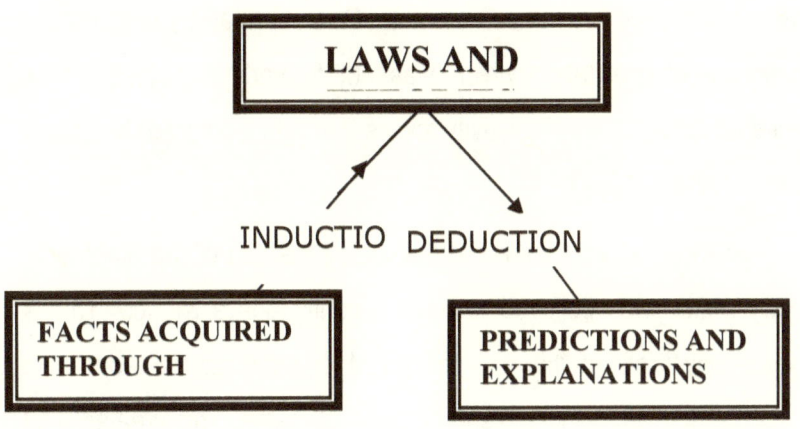

The above schema gave room for Nicholas Copernicus to formulate a comprehensive heliocentric approach of the universe which displaced the Ptolemaic cosmology of the earth as the centre of the universe (geocentricism). Copernicanism had influenced the works of Isaac Newton to formulate his laws (Mendie 125).

The laws of Newton served as a base or a foundation for science to grow in the 17th to 19th century cosmology. It was in between this time that the speed of light was discovered in 1876 by Danish astronomer Ole Roemer. It was this period that the Discovery of the theory of relativity was made by Albert Einstein in 1905. Within this thought system Albert Einstein was forced to give a new meaning to the entire discoveries of Newton. According to Einstein in his Special and General relativity, space, time and momentum are not absolute but a product relative to the observer's frame of reference (Essien, Ephraim, 2007). In 1905 Einstein discovered the special relativity which was based on "observer's frame of reference". Thus, he posits that, time and space is relative in relation to the observer's frame of reference. In 1916 he published his famous work on General relativity which takes cognizance of gravitation which was not formerly explained in the special relativity. General relativity shows that light is affected by gravity (Einstein, Albert, 1984).

However, it was observed that up till the end of the nineteenth century matter was believed to have a continuous existence. This is why the physicists of this era conceived that the natural world is a continuum, and simply in a blend of forms of matter. This created a lot of problems for the classical physicist. The classical theory is unable to explain experimental observations regarding small object which exist in the sub-atomic realm of reality. They could not give explanations concerning the activities of the sub-atomic world. Others include:

v.   Discrete spectra emitted by excited atoms.

vi.   Photoelectric effect

vii. Variation of heat capacity of mono-atomic solid with temperature

viii.   Spectral distribution of energy in black body radiation etc.
(Shashi Q C, 2010:1).

With the above problems there was need for a new theory called Quantum mechanics.

THE DEVELOPMENT IN QUANTUM MECHANICS: THE ROAD TO STRING THEORY

In attempt to solve the above problems, classical physics was to give way to this new conception of nature and so, a new physics called quantum mechanics was born. According to Gary Zukau (The Dancing Wuli master), he defines quantum and quantum mechanics as follows:

> A "quantum" is a quantity of something, a specific amount. "Mechanics" is the study of motion. Therefore, "Quantum mechanics" is the study of the motion of quantities. Quantum theory says that nature comes in bit and pieces (quanta) and quantum mechanics is the study of this phenomenon (45).

This radical change from the longstanding view about nature occurred in the beginning of the twentieth century paving way for increased rationalism in science. The radical view was that nature was not simply a blend of forms of matter or physical quantities that has a smooth, continuous existence. Rather, nature was discontinuous and discrete. It is a new dimension of science that evolved during the first new decades of the 20th century in an attempt to understand the fundamental properties of which matter is made up of. It started with the study of the interaction of radiation and matter, with the use of classical mechanics, radiation effect could neither be explained, nor the theory of electromagnetism. In particular, physicists were puzzled by the nature of light's peculiar lines in the spectrum of sunlight which was discovered earlier by Joseph von Fraunhofer (1787-1826). These spectral lines were systematically catalogued for various substances, yet nobody could explain why the spectral lines are there and why they could differ for each substance. It took about one hundred years, until a plausible explanation was supplied by quantum theory.

In a contrast examination to Einstein's relativity, Newtonian physics is deals with larger objects in the universe, while quantum mechanics or quantum theory deals with the tiniest things we know, that is, the particles that atoms are made of, which is call "subatomic particles". It is imperative to note that quantum mechanics was not the work of one individual, but the collaborative effort of some of the most brilliant physicists of the 20th century, among them are, Niels Bohr, Erwin Schrödinger, Albert Einstein, Wolfang Pauli, Max Born, Max Planck (1858-1947) and Werner Heisenberg (1901-1976) it has been on record that, Planck is the originator of the quantum theory, while Heisenberg formulated one of the most eminent laws of quantum mechanics, which is the "uncertainty principle" , it is occasionally referred to as the "Principle of indeterminacy".

Max Planck's remarkable contribution was that "Energy is not continuous" which is referred to "Planck's constant". It all happened around 1900 at the university of Kiel were he concerned himself with observation to the radiation of heated materials in attempt to draw the conclusion from the radiation to the radiating atom on the basis of empirical data, he developed a new formula which later showed remarkable agreement with accurate measurement of the spectrum to heat radiation. The result of this formula was so that energy is always emitted or absorbed in discrete units, which he called quanta (Hawking 1988:83). He further developed his quantum theory to derive a universal constant, which came to be known as "Planck constant". The resulting law states that energy of each quantum is equal to the frequency of the radiation multiplied by the universal constant E=hv. The discovery of quanta revolutionized physics, because it contradicted conventional ideas about the nature of radiation and energy. Planck's hypothesis that energy is radiated and absorbed in discrete "quanta" X (or energy element) precisely matched the observed patterns of blackbody radiation (Planck, Max, 1967).

According to Plank each energy element E is proportional to its frequency v: E =hv (Ayi, 2006:405) where h is Planck's constant. Thus, the 1905 Albert Einstein interpretation of Planck's quantum hypothesis realistically was used to explain the photoelectric effect in which shining light on certain material can eject electron from the material. Albert Einstein's serious contribution further shows that an electromagnetic wave such as light could be described as particle (later called the photon) and this was established in 1926 through the work of Bohr and Heisenberg. They published results that closed the "Old Quantum theory", which explained that discrete quantum energy of photon is independently on its frequency. According to Egbia (2010); modern inventions such as television, camera tube, solar power cell and photographic light meters all relate to Einstein's description of the photoelectric effect (14). Hence further development of unity theory between the atomic particle and electromagnetic waves called wave-particle duality, in which particles and waves were neither one nor the other, but had certain properties of both was accomplished. Quantum mechanics described the world of the very small; it explained and investigated macroscopic system such as superconductors and super fluids which classical mechanics could not explain. Quantum mechanics through its modus Operandi and experiment, repeatedly produced result which the physics of Newton could neither predict nor explain. Although Newton's physics could not account for phenomena in the microscopic realm, it however explains events in the macro realm. Quantum mechanics is based upon experiment conducted in the subatomic realm. As Zukau may assert, it predicts probabilities, these probabilities are subatomic phenomena. Subatomic phenomena cannot be observed directly, none of our senses can detect them, not even any scientific apparatus can tell us exactly its true state, because the subatomic world keeps changing and the apparatus itself keeps defecting its true nature. It is pertinent to note that not only has no one ever touched an atom much less seen an atom, much less an electron, no one has ever tasted, touched, heard, or smelled an ether. This is what quantum mechanics can explain. It is the new dimension of science; Quantum mechanics claims that the deterministic view of reality does not hold in the quantum world. The atom seen as the building blocks by the classical physicists has been split into further minute particles like quarks, spins, photons and leptons, mesons, etc, which is said to be more than a million times smaller than the head of an office pin (Akpan, 2005). Thus, neither the movements nor the position of the subatomic particles can be measured or predicted with certainty. However, it is observed that what holds at this level is randomness and chaos, which can only be explained by the "uncertainty principle" or the "indeterminacy principle", hence only probabilities are precisely determined in quantum world. This makes quantum theory, or quantum mechanics to shatter the three basic assumptions of classical science which include:

iv. Causality

v. Determinism

vi. Objectivism.

It is pertinent to note that atomic particles are the ultimate materials from which all the complexity of existence in the universe emerges; they do not obey deterministic and predictive laws of Isaac Newton, hence their activities are causally indeterminate and could only be understood with probability.

However, classical mechanics works very well at this present time for large objects that are moving slower than the speed or velocity of light. On the other hand, when objects are so small, quantum theory is needed for a better explanation. This is why in quantum mechanics, some of these behaviors are microscopic and only emerge at extreme low or very high temperatures or energies. Hence, Quantum mechanics is derived from the observation that some physical quantities can only change in discrete amount called "quanta" and not in a continuous analog way. For example, the angular momentum of an electron fastened to an atom is quantized.

In the content of Quantum mechanics, the wave-particle quality of energy and matter and the uncertainty principle provide a unified view of the behaviour of photons, electrons, and other atomic-scale objects as formulated by Rutherford which classical physics could not. Many of the results of quantum mechanics are not easily visualized in terms of classical mechanics. For instance, the "ground state" in a quantum mechanical model is a non-zero energy state that is the lowest permitted energy state of a system, as opposed a more "traditional" system that is thought of as simply being at rest, with zero kinetic energy instead of a traditional static, unchanging zero state. According to John Wheeler in his essay entitled *spooky quantum;* he contends that quantum mechanics allows for far more dynamic chaotic possibilities that cannot be seen by our senses. Thus, quantum mechanics incorporates four classes of phenomena for which classical physics cannot account:

vii. The quantization of certain physical properties

viii. Wave-particle duality

ix. The uncertainty principle

x.   Quantum entanglement

xi. Photoelectric effect

xii. Black body radiation. Etc.

Schematically, the following are the important contributors of the foundation of quantum mechanics and principles they uncovered (Mendie 125).

YEAR	RESEARCHER	CONTRIBUTION
1901	MAX PLANCK	Blackbody Radiation
1905	ALBERT EINSTEIN	Photoelectric effect
1913	NEILS BOHR	Spectra theory
1922	COMPTON	Photo scattering
1924	WOLFANG PAUL	Exclusion principle
1925	De BROGLIE	Matter wave
1926	ERWIN SCHRODINGER	Wave equation
1927	WERNER HEISENBERG	Uncertainty Principle
1927	DAVISON & GERMER	Wave properties electrons

However, Quantum mechanics had enormous success in explaining many of the features of our world. The individual behaviour of the subatomic particles that make up all forms of matter (electrons, quarks, protons, neutrons, photons, spins, leptons, mesons, etc) can often only be satisfactorily described using quantum mechanics; hence the interpretation of quantum mechanics at the city of Copenhagen.

The Copenhagen interpretation of quantum mechanics accepted the probabilistic nature of quantum theory which for them is not a temporary feature which will eventually be replaced by a deterministic theory, but instead must be considered a final renunciation of the classical idea of causality. It is also believed therein that any well-defined application of the quantum mechanical formalism must always make reference to the experimental arrangement, due to the complementary nature of evidence obtained under different experimental situations. They accepted firmly on the uncertainty principle of Heisenberg which explains that in determining a particle or an electron moving in an orbit, that both "position" and "momentum" cannot simultaneous be measured with complete precision. However, one can measure the position (alone) of a moving free particle creating an eigenstate of position (alone) of a moving particle, creating an eigenstate of position with wave function that is very large at a particular position x, and zero everywhere else. The Copenhagen team adopted that quantum mechanics as the last theory science could ever achieve.

This adoption excluded Einstein's relativity and became a problem which Einstein fought till his dead but could not solve. He rather searches for a "unified field theory" which could marry relativity and quantum mechanics together called string theory or the theory of everything.

WHAT IS STRING THEORY?

String theory currently is the only theory vying for the total explanation of all known entities, the world of matter, theory of relativity and quantum mechanics. This theory predicts that all forces of the universe could be unified into a single manipulation of tiny loops of vibrating energy called strings. It all started, when Einstein tried to unify relativity with quantum mechanics together in his unified field theory. The quest for a "unified field theory" which could give room for the establishment of an underlying link between the seemingly unrelated force of gravity and electromagnetism initiated by Einstein was to solve the problem of the vision of marring general theory of relativity with sub-microscopic quantum world of quantum mechanics is what string theory attempts to offer.

In the works of John D. Thouless(1961), he posited that Einstein had recognized that this new quantum theory enjoyed remarkable empirical successes. However, he did not believe that future fundamental physics should be built upon it. Rather, he thought the way ahead was to develop the geometrical approach of general relativity into an all encompassing "unified field theory" within which the results of the new quantum theory would be derived.

String theory posits that every matter, momentum, force, subatomic particles are reducible and could be seen as tiny vibrating loops of energy called strings. Meaning that, the entire universe including man is made up of billions of strings that hold things together. And these strings can relate so easily with all the dimensions of the universe into one single dimension of which the strings are form. When a string comes together with another, matter is formed, and when they go apart, matter goes out of existence. This is the only theory vying for the unification of all reality, as well as the theories explaining reality, including relativity and quantum mechanics. But the question is, is string theory real? Can it be proven?. Here we shall consider the position of the realists and the idealist in support or against the existence of string theory.

THE CONCEPT OF REALISM AND IDEALISM: AN EXPOSITION

Realism as a philosophical school of thought suggests that reality exists independent of the workings of the human mind. In other-words, all realities exist, and are not in any of the constructs of the human mind. For instance; a Tree exists, and it presents its existence to me. Whether I perceive that Tree or not, does not in any way influence or counteract its existence. Its existence is independent of my ratiocinative thought. In realism objects speak for themselves without our judgments on them (Uduigwomen, 2007:149). This view is in direct contrast with idealism. Idealism as a school of thought, which simply opines that realities are basically constructs of the human mind. Nothing can exist independently of the human mind. A Tree exists only because I think of it and extend the categories of my mind in perceiving it. Thus for the idealist, all realities are basically constructs of the human mind and as such, reality is dependent on the human mind. Anything that exist, according to the idealist are products of the mind, they are created and categorized by the mind before one can perceive them.

THE IDEALIST AND REALIST DEBATE ON STRING THEORY

String theory in itself presents a philosophical unique problem when we make attempts to decipher whether it is idealism or realism. This problem is quickly brought to the fore when we discover that string theory itself remains unproven at this moment. The theory in itself was promulgated as an explanation to solve both the existing disharmony existing when we take into cognizance the major theories in physics via quantum mechanics and the theory of relativity, as well as giving an all embracing explanation of the underlying principles and basic component of nature. With its basic foundations basically idealistic in nature, one can easily aver that string theory remains an idealistic construct. However, if we delve deeper, we realize that this theory logically follows and is based on a sound mathematical foundation. Mathematics in itself explains reality in an abstract form. If this is so, what is implied is that string theory by extension also explains reality.

The conundrum thus remains. Do we consider string theory a realism or idealism? This paper suggest that string theory is indeed an idealistic construct capable of solving scientific problems if properly utilize. But the question is can string theory really explain all the forces of nature?, can it really capture everything in the universe?. These are some of the thought provoking questions philosophers and scientists should consider when examining the faith of string theory as the theory of everything. According to the idealist all that exists are ideas constructed by the mind, and as such since String theory captures that, then its existence is true. This paper supports the position of the idealist because most time what we claim to see with our real eyes do fails us, in other words our senses are not perfect at all point (Ozumba, 2001:43)

CONCLUSION

The success of explicit explanations of interactions in elementary particles physics saw an unprecedented expansion of our understanding of the physical world, was remarkable in twentieth century physics. This gave rise to the study of tiny loops of vibrating strings of energy called strings theory. This theory was to replace the quest for the theory of everything, which could unify classical mechanics, theory of relativity and quantum mechanics into one unified reality, a reality that is ruled by tiny loops of vibrating energy called strings. We have shown that the thesis adopted by the idealist occurs to be a better stand and can give science the future it has been seeking for, a future whereby the problem of the universe can be given adequate attention. In view of the paper it proves and shows that science is a continuum, one should not be in a hurry to abandon an idea because it has not been proven empirically, thus the hope of being proven in the future is a possibility.

CHAPTER TEN

RATIO-EMPIRICO-CENTRIC METHOD: A CRITIQUE OF EMPIRICISM AND RATIONALISM IN MODERN SCIENCE

Introduction

Science within and before the 19th century was driven mechanistically, its operations were defined as an experiential discipline; observation and all other faculties of sense data collections were the key factors of its greatest achievements. These empiricist doctrines became the determinants of what should be called science in the modern age, influenced by logical positivism as the building block of all scientific justification. This mechanistic understanding of science has influenced a lot of scholars and modern thinkers up-till today, making them to conceive science as the handwork of empiricism alone. This essay, therefore, is an attempt to excavate moments in science which have shown rationalist contents of science rather than that of the empiricist alone; The work shall insightfully examine moments of formation of theory of relativity, quantum mechanics and unified field theory (string theory) as the fore-wheel that is today the building force of science, which has become part of postmodernists contents in science. The work explores the role of postmodernism, backed by the philosophical method of critical analysis, logical argumentation, evaluation and prescription. The work posits that, in view of the above theories, science should not be seen as an empirical discipline alone neither should it be seen as a rational discipline alone because of their limitations. As such, there should be a rejuvenation and revitalization of our thought system through a postmodernist style of thinking to embrace the postmodernist and the empiricist contents of science, which have proven to be the most essential method of science and technological development in our contemporary age. In evaluating this work, the work adopted the postmodernist approach of ratio-empirico-centric method in accessing, evaluating and analysing the various approaches of modern science towards a complementary standpoint of understanding the way nature manifests in the study of science.

Conceptual Clarification

Human beings by nature are naturally faced with the drive to aptly investigate and to give meanings to the general composition of the universe. Most times, if one is asked to explain the composition of the universe, one will be tempted to make a list of stars, moon. Sun, oceans, rivers, buildings, air, momentum, space, time and so on. These are some elements of reality that can be sensed directly through human sense data collections via sense organs. This direction of mechanistic understanding of the universe is what coloured Galileo, Isaac Newton, and so many others to picture the universe as a product of the five senses; the senses of sight, sound, touch, smell and taste.

The outcome of Newtonian physics was a product of these factors as the building block of science. Thus, the laws of Newton were based on that understanding of the universe as mechanistic. The science of Isaac Newton introduced us to a modern approach of how the laws of nature operate, and the mystery behind the entire universe, giving us a deeper meaning about reality parting from the mythological approach of the pre-ancient philosophers.

The Galilean and/or Newtonian science consider the empirical or observational explanatory model to justify the ultimate reality. For this reason causality was a vital force to all scientific investigations because there was certainty, and the science of that era was with absolute determinism. In other words, for any reason that an effect is observed there must be a cause and every cause has its effect in line with the principles or laws of classical mechanics as by product of the empirical tradition.

However, a deep reflection to the world view of Newtonian science which was the ground base of modern science, one could discover how empirical explanatory model became the driving force of science, prediction also was the handmade of casualty which were all predominantly coloured by induction and justification of fact deduced from the mechanistic laws of nature (Chalmers, 8).

The notions of empiricism is what possibly directed the heliocentric approach of universe to replace the Ptolemaic cosmology by Nicholas Copernicus and also a product of what gave Isaac Newton the driving force to find certainty in knowledge through experimentation. This attitude coloured the hallmark ideas and principles adopted by the logical positivist tradition to adopt the verifiability criterion as the only scientific way of studying the universe. For this reason empiricism thrived for so many years in science and giving meaning to only those factors that could be directly experienced or would possibly be experienced in the future.

But this ideology about nature (as mechanistic) was first overturned by a great physicist called Albert Einstein. He gave us a new insight of how we should understand the laws of the universe, a universe that does not operate from an absolute space or absolute time; a universe driven by the observer's frame of reference. This new picture of the universe is what he expressed in his special and general theories of relativity in 1905 & 1915 that gave birth to a modern rationalist content of science. Special relativity informs us on the rationalist role of space and time as relative but one that cannot be separated, and general relativity explains the rationality in a gravitational field both from the observer's frame of reference (Gribanov, 217). And from this moment rationalism began to play a dominant role in physics; and other science related disciplines started adopting the ideas of Einstein as an innovative view that overshadowed absolute classical laws of Isaac Newton. This also coloured the later theory of Einstein called the principle of photo-electric effect, which states that a photon with high dense energy when contacted with a metal of low dense energy, the metal will emit photo-electrons which can be used to produce electricity. This aspect of Einstein work shows the rational power of the German Physicist.

In no distant time, the guiding principles of Einstein relativity and Newtonian physics could no more solve some of the looming questions of science. It was there to solve problems of the macro world, and could not study any defining moment in the micro world. Thus, a new science was needed; through Marx Planck's action of "quanta" in 1905 a new science called quantum mechanics was developed to give meaning and explanation to sub-atomic particles of the micro world which was not accounted for by relativity and even by classical laws of Isaac Newton. Quantum mechanics pursued the aspect of rationalism in science without paying attention to what we can experience; by this it investigates the unobservable part of nature, the world of quarks, spins, leptons, photons, mesons, plasmas, electrons and protons, et cetera.

However through the light of postmodernist understanding, Einstein a contributor of quantum theory attempted to marry his relativity theory with quantum mechanics to give a more explicit understanding and a unification of all laws of the universe which shows the rationalist notion of quantum gravity; an idea that is today developed as string theory. String theory is the only existing theory that is vying for the total explanation of all entities (elementary particles) world of matter, theory of relativity and quantum mechanics. Its prediction is so powerful, fundamentally stating that everything in the universe is reducible to strings. Strings are tiny loops of vibrating energy surrounding the entire universe which can come together to form matter. String theory predicts that everything about nature including man, stone, animal and many other things one can find in the natural universe is made up of these tiny vibrating loops of energy called strings.

These three moments of scientific development and revolution have shown us that empiricism in science alone has limitations on the way the universe operates. Hence, the driving force of science has departed from viewing science as an empirical discipline alone. Therefore, theory of relativity, quantum mechanics and string theory have sharpened our understanding of the entire universe and has given us a new picture of science which are rationally based and also the understanding that nothing should be absolute as pursued by postmodernism. Thus, the impetus of this research is captured from this background highlighting that these moments in science have a remarkable impact in technological development which are all founded on the principle of rationality and styled in a postmodernist way. But the work posit further that, empiricist or rationalist method of science in their state of singularity cannot solve all scientific problems because of their limitations, this is why this work seeks for a ratio-empirico-centric method, a system that will complement both dimensions for an holistic and complementary approach in seeking scientific knowledge that is geared towards unity in diversities, a 'ratio-empirico' complementation in the acquisition of scientific knowledge based on the contexts of the scientists.

Empiricism and its limitations in Modern Science: A Postmodernist Critique

Modern science is an era of scientific investigation that was dominated by observation and experiment. This tradition is unique by its approach, on a systematic cum experiential study of the entire structure of the universe.

A . F. Uduigwomen in his thought provoking book gave us a vivid illustration of the nature and scope of empiricists' notion of science within modern science. Science for modern scientists according to Uduigwomen is a kind of knowledge arranged in an organized or orderly manner, especially knowledge derived from experience, observation and experimentation (*A Textbook of History and Philosophy of Science*, 20).

The exposition by the author above has given us the background on the shape of modern science and their method of research. Science at this age was totally influenced by the works of Galileo and Isaac Newton. Both scholars were indeed chief advocates of modern systematic thinking. The era of Isaac Newton was remarkable, and fashioned after objectivity, causality and determinism of all scientific knowledge. These three assumptions in modern science helped to systematized and compel scientists to a remarkable standard all over the entire world. Isaac Newton reduced science into precise mathematical laws capable of pushing in objectivity in science. These laws explain everything about the world of matter (Macro world) and everything ever seen using sense perception; where all are controlled by the laws of Newton. These laws are what Newton called the three Laws of motion and the law of gravity.

Isaac Newton (1642-1727) revolutionized physics with his proposition that all bodies are governed by the three laws of motion which he used in describing the motion of the planets and the moon. Let us consider his three laws as posited by John Gordon (et al):

4. The first law of motion (also called the **law of inertia**) states that an object at rest will continue to rest until moved by another object and an object in motion will remain in motion with a constant velocity unless acted on by a **net external force.**

5. The second law (also called **the law of acceleration**) states that the force applied to an object is proportionate to its mass multiplied by acceleration (F = Ma). Meaning the acceleration of an object is directly proportional to the net force acting on it and inversely proportional to its mass
6. The third law (also called **law of action-reaction**) states that for every action there is an equal opposite reaction. (*Principles of Physics*, 64)

Thus, with these three simple laws, Newton created a whole new model of the universe, superseding Ptolemy's model of epicycles; who eighty years before Galileo (1564-1642) had pointed out that the earth rotates around the sun, and brought in a mechanistic view about the universe. This Mechanistic view about the universe by Newton and Galileo provided the basic for 17th to 19th century cosmology, mechanistic view that sees the universe as an arrangement in which stars and planets interact with each other like springs and cogs in clockwork manner. Thus, for the initial positions and states of all objects in a mechanically determined universe are known, all events can be predicted until the end of time, simply by applying the laws of Newton. The mechanistic view of the universe brought about new hypothesis of interpreting reality determinism. Hence, underlying determinism is the view that everything in the universe, including all the motions, from the smallest to the largest occur in a way that can be predicted with absolute accuracy using the laws of Newton, nothing is left to chance (Pagels, 4). It is quite imperative that one cannot discus Empiricism in Modern science without having a critical survey of the contribution of Isaac Newton, and Galileo because they were the greatest figure that really shapes scientists into modern light. This is why many scholars regard Isaac Newton as the father of modern science.

In line with Pagel's exposition, Akpan on *Quantum Mechanics and the question of Determinism in Science* contends that:

> Classical science and in fact post-Newtonian science up till the early twentieth century were mired in a deterministic interpretation of realities. The deterministic hypothesis in science holds that everything in nature has a cause and if one

could know the antecedent causes, he could predict the future with certainty (Sophia: A Journal of Philosophy,72).

The above explains that, in the classical physics, determinacy and prediction were possible. Gary Zukau, on *The Dancing Wuli Masters: An Overview of the New Physics* rightly agree on the predictive powers of classical mechanics and opines that:

> Newton's laws of motion describe what happened to a moving object. Once we know the laws of motion we can predict the future of a moving object provided that we know certain things about it initially. The more initial information that we have, the more accurate our predictions will be....for example, if we know the present position and velocity of the earth, the moon, and the sun, we can predict where the earth will be in relation to the moon and the sun at any particular time in the future giving us a foreknowledge of eclipses, seasons, and so on (50).

He further states that:

> According to the old physics, however, it is possible, in principle, to predict exactly how a given event is going to unfold, if we have enough information about it..... The ability to predict the future based on knowledge of the present and the laws of motion gave our ancestors a power they had never known (51).

Gary Zukav's work gave us the background of the predictive power of classical mechanics using Newton's laws full of certainty.

Conversely, empiricism in science up till today has suffered a lot of criticism both from scientists and philosophers of science because of its enormous limitations. The questions that may puzzle critical minds are; can only the observable be studied as scientific? Can observation alone solve all human problems in science? How certain can observation inform us about the true nature of things? These are few of the thought provoking questions that a philosopher of science may be tempted to ask about empiricism in science.

In reaction to these questions, science from Feyerabend's perspective can be investigated without the help of experience, because for him theories are just strings of signs without relation to the external world, unless we now design mechanism of connecting them to experience, experience will not be necessary. Testing of fact may not need experience before one can test a particular theory; it is the work of human ratiocination to communicate on the process of testing and to even understand the contents of a particular theory after testing. According to Feyerabend, it is easily seen that experience is needed at none of the three points listed above (791). In his criticism, he further noted that experience arises together with theoretical assumptions, not before them, and that our experience about the natural universe without theories is just as uncomprehended as is a theory without experience. This is what he termed as observational-theoretical dichotomy. The information is that, for Feyerabend observation should not come before theoretical knowledge of a given reality; as such theories which are rationally based should front our notion of science, because knowledge for him can enter our brain without torching our senses. The question here is how? He submitted that, the mind has the power to create knowledge through deduction of ratiocinative reflections that do not connect with experience which may be gotten through intuition.

However, in view of the moments of formations of relativity by Albert Einstein, formation of quantum mechanics through Max Planck discovery of Quanta on his Blackbody experiment, and moment of string theory; one can insightfully capture that empiricism in science has a lot of deficiency. Because of the limitation of empiricism in classical mechanics, Shashi Chawla agreed that, the classical theory was unable to explain the following experimental observations regarding small objects; which includes the following:

iv. Discrete spectra emitted by excited atoms
v. Photoelectric effect
vi. Variation of heat capacity of mono-atomic solids with temperature and, Spectral distribution of energy in blackbody radiation etc. (*A Text Book of Engineering Chemistry*. I).

Meaning empiricism became less reliable in science, scientists needed to move into a postmodernist way in other to solve their problems, and they had to part ways to quantum mechanics where rationalism reigns and where the problems of subatomic particles could be explained. This is why, in this research we attempts to fashion a new paradigm for science that will embrace both methods of science (empiricist and rationalist methods) from an integrative standpoint. Let us proceed by exposing the rationalist view of science and its limitations.

Rationalism and its limitations in Modern Science: A Postmodernist Critique

This section centres on rationalism in Modern science, its critique and finally the limitations of rationalism in Modern science. Modern science did not celebrate greatly on much of rationalist contents; because the idea of observation, experimentation and verifiability criteria of the logical positivists coloured the entire gamut of modern ideology about the method of science. But this research is tasked to excavate some rationalists' element of that age that can be said to be devoid of empiricist notions. Such moments include; moments of formation of relativity, quantum theory, and string theory which was then called the theory of everything. Thus, modern rationalists are of the view that all scientific knowledge starts from innate ideas, as such through reason one can gain an authentic ratiocinative knowledge of the nature, scope and origin of the universe.

However, there are **fundamental tenets of rationalism in modern science** which includes:

5. Reason is the only authentic medium of scientific knowledge.
6. The power of science is only derivable through ratiocination and not experience or observation.
7. The method of science is the method of deduction.
8. All knowledge comes to man through innate impulses and are justified through rationality.

The above fundamental tenets of rationalism were highly operational within the studies of relativity physics, quantum mechanics, and string theory and also suggest the place of subjectivity in science. Subjectivity is a tradition of belief that explains that the underlining factors of the universe can only be discovered from an individual standpoint. It lends credence to the value of personal fancies in the studies of science. Through reason, the radical rationalist opposed all knowledge claims that are provable by experience or observation; thus the only way to understand reality is not by mechanistic study of the universe rather by logical cum deductive work done through the aids of reason mostly reflected in relativity, quantum mechanics and string theory.

However, how does rationalism manifest in quantum mechanics? We noted that earlier that quantum mechanics is built under the caprice of rationalism. Because, scientists cum philosophers of science needs the help of reason to understand quantum theory; thus, the science of quantum mechanics studies the unobservable part of reality. Never on earth has a scientist heard the sound of a photon, or felt the weight of mesons, neither has one ever torched an electron or photon. This part of physics can only be studied by uncertainty and indeterminacy principle. A principle stated by Niels Bohr and Werner Heisenberg that the more précise you determine the position of a subatomic particles the less precise the momentum can be. These inform us the rationality behind quantum mechanics that has revolutionized modern science. This theory shattered the foundation of classical mechanics by eradicating absolutism in physics because at the level of quantum mechanics the deterministic laws of Newton ceases to exist.

String theory in science is another important moment that has given us a new paradigm of rationalism in science. According to the theory, reality can be reduced into some tiny loops of vibrating energy called strings. String theory currently is the only theory vying for the total explanation of all known entities, the world of matter, theory of relativity and quantum mechanics. This theory predicts that all forces of the universe could be unified into a single manipulation of tiny loops of vibrating energy called strings. The long quest for this theory has been a demanding development to search for the theory of everything by modern physicists which started with the works of Albert Einstein to unify his theory of relativity with quantum mechanics called quantum gravity that made it possible to view reality as one dimensional string. This theory predicts the existence of dark energy, dark matter and subdue everything in the universe into tiny vibrating unseen strings. Thus, its reality remains uncertain but has been proven to be in existence only through the aids of mathematics. This has proved the notion by Feyerabend earlier noted that science can be done without the aids of experience or observation (Science without Experience, 794). Therefore, it is true that all these moments of quantum mechanics and string theory are moments in science that neglect observation and experience. Conversely, the idea of rationalism in modern science has enormous criticism and limitations, let us quickly establish these limitations and to further establish a new paradigm of thought called 'ratio-empirico-centric' method.

The limitation of rationalism in modern science shows that there is no scientific or philosophical work that may lack criticism; this is because no work can be free from certain limitations. The rationalist view in modern science have been criticised by many scholars globally. Firstly, in a work published by Einstein, Podolsky and Rossen (EPR) in 1935 entitled *Can Quantum Explanations of Physical Reality be Considered as Complete?* In this work Einstein and his colleagues argued that the idea fronted by quantum mechanics cannot be said to be a complete explanation of the entire reality. They did not agree to the position Bohr and Heisenberg claimed about quantum mechanics. Though, Einstein had contributed in the development of quantum mechanics on "photoelectric effect". In reaction to this, Einstein spent a large portion of his career arguing against quantum mechanics as the last theory of science, because for Einstein a theory that can only be explained by mere rationalism without experience or observation cannot be the last theory in science. Einstein and his colleagues conveyed a message that, what is appearing as indeterminacy and uncertainty, is because we do not know the real nature of reality, so because we don't know it, it appears uncertain and it appears indeterminate, immediately we know the deeper reality, it will reveal that; what is looking as uncertain or indeterminate is because of our lack of knowledge, because "God does not play Dice".

In relativistic physics, the concept of black-hole which was first predicted by general relativity, asserts that there is an infinite hole absolutely black, absorbing light and does not emit out any energy. Thus, in extreme gravity in the black hole, Einstein's rational equation seems not to contain; at that level all laws of physics breaks down to infinite; and infinity in mathematics is a number without limits, but in physics it is the collapse of everything about the physical universe. Therefore, there is a problem with Einstein assertion of the black-hole. This led to what modern physicist called the singularity of the black-hole, meaning scientists can't predict what is going on or the possible outcome of the black-hole. To solve this limitation of the black-hole, quantum mechanics was not enough, meaning there was need to combine gravity to quantum mechanics called quantum gravity (in modern physics string theory). Another problem emerged, the problem of infinite sequence of infinity. This combination failed because quantum mechanics and theory of relativity failed apart, this led to the collapse of physics. A problem yet to be solved even with all tools of rationalism in science, in other words, nature is smarter than what we know through rationalism alone. The methods of rationalism failed to capture quantum gravity; meaning reason alone has enormous limitations.

In view of the above exposition, we notice that rationalism alone as the method of science cannot lead us to unity in science, as such to solve this dichotomy between rationalism and empiricism in science there is need for another approach in science that could take credence of both traditions. This is why we have coined a new word to present a new paradigm of science called 'ratio-empirico-centric' method. The work applies this system of philosophical approach in science as a new paradigm for integrating both traditions of science.

Rejuvenation and Revitalization of Scientific Methods through Ratio-Empirico-Centric Method: A Postmodernist Evaluation

The word **'Ratio'** is rooted to the school and the ideology of rationalism that project the role of reason or innate ideas as an authentic means in knowledge acquisition. **'Empirico'** represents the school of empiricism that emphasized on experiential means of acquiring knowledge; meaning human beings were born as a blank slate but through experience a true knowledge is acquired. And **'Centric'** here implies 'Knowledge that is centred on something'. It is a Philosophy because, it an epistemic approaches that critically evaluates into the traditions of rationalism and empiricism for a full understanding of the origin, nature and scope of the entire universe and how it manifest to human beings.

What is Ratio-Empirico-Centric Method

Ratio-Empirico-Centric method is a postmodernist way of studying science from both the rationalist and empiricist dimensions in a postmodernist way. Because the postmodernist believes that there is no one way to study reality, therefore both ways are imperative. It is an integrative cum complementary approach of studying science from both views of rationalism and empiricism. It shows that knowledge begins with reason and could also start with experience depending on the contexts of the scientist, which later manifests in reason or experience as the case may be. Thus, it projects an ideology that in uncovering the mystery behind nature and how the universe operates within the moments of formation of theory of relativity, quantum mechanics and string theory, that reason through to tools of the mind captures reality first and later proceeds information to the senses organs for fact verification with moves back to the brain for onward processing. And within the empiricist tradition knowledge may also start from experience and later justified through the aids of reason. Here, the sense organ may capture images through the eyes and send it to the brain for reasoning to take place.

Within this understanding, the philosophy provides a background that science operates in such an integrative mood; were the method of science may either start from ratiocination stage and later gets confirmation through experience and vice versa. The philosophy in another way, is essential when dealing with elementary particles within quantum mechanics and string theory (Particle Physics) because as particle element they are unobserved, but they exist to the eye through (particles formation) when they come together to form matter which could then be seen through experience. Because of this background, this research seeks for rejuvenation and revitalization of our thought system to adopt both methods of science as a necessary and essential way of understanding the nature, scope and properties of the entire universe; because reality is better understood from the various contextual ways it appears to us. In other words, there is no absolute way (objective point) one can grab reality, this should be applicable to science through this philosophy; science should not be seen as the handmade of empiricism or rationalism alone, but from any way it appears to the scientist in search for a particular finding he or she wants to investigate on. For this new paradigm to be sustained; there is greater need for cultural regeneration, where everyone will show evidence of resurgence through adoption of this new paradigm in our new model of thinking about science and in action by adopting rationalism or empiricism in science from any contextual point the scientist finds suitable.

The pillar of science in contemporary view is rested on the conjugation of both empirical and rational approaches to science. In other words, due to the limitations of both methods of science in cognition of true reality; there is need to rejuvenate and revitalise science through the method of ratio-empirico-centric method. In physics or entire sciences, Isaac Newton drew out a clock work universe that helps to solve most of the challenging issues in science with certainty, and thus, help to defined science mechanistically using the collection of his senses and experimentation. His laws of motion and gravitation all proved enormous success in physics and mathematics (calculus), and philosophically proved that absolutism is another way nature can manifest through fixed laws based on observation and our experience of the universe. But theory of relativity, quantum mechanics and string theory have come to create another unique way of understanding the universe, were reason through the tools of consciousness can create reality outside experience called qualia (Deepak & Menas, 287).

But this work states that, to understand reality in an holistic dimension, all manifestations of the universe from both rational and empirical strains must be conjugated, meaning the visible and invisible, subjective and objective, physical and non physical, experiential and non-experiential must be studied as part of science that should make up universe. This is because; in the formation of a complete universe there is a complementation of all existing factors, this is why Asouzu describes reality as the complementation of all missing link; missing links here are the various manifestations of reality in units, parts and in different compartments (Method and Principles 285-286).

Thus, having a close look at the various manifestations of reality through ratio-empirico-centric method, if two people are observing an aqua-coloured vase, they may choose to disagree over whether the colour is green or blue, but will not disagree on what "colour" means. They may also taste a bitter kola and argue over whether it is bitter or not, but they won't disagree that taste is involved; they may also observe the dominant colour of Nigeria national flag, here they may argue whether it is green or white. In another case, they may also choose to observe the behaviour of light, one may say it appears like a particle and the other may argue that light is actually a wave. Here both views are correct because from the background of wave-particle duality of light, water, sound and any others, quantum mechanics has proven that everything in the universe can be reducible to particle-like properties or wave-like properties but not simultaneously. This was proven in 1923 by Authur Compton who experimented by firing at X-rays on electrons, previously X-rays were known to be waves but Compton in his thought experiment proved that X-rays can both act as wave and also as particles, this experiment was called *Compton scattering experiment*. However, we shall describe the equivalent experiment done with X-ray and photons, which as we see bellow, how the experiment was eventually carried out definitively.

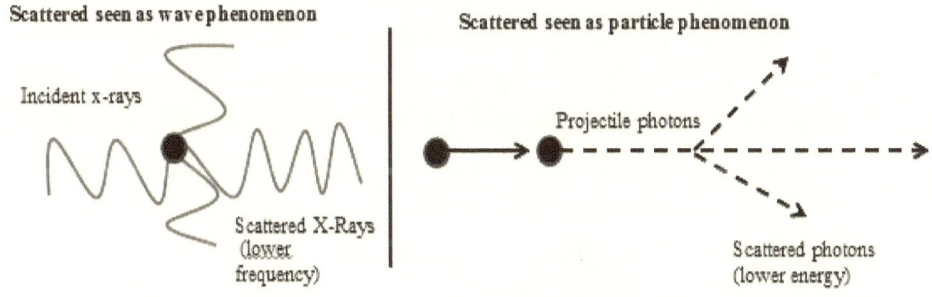

Compton scattering experiment

However, for these two observers, they need to complement their rational and empirical knowledge to understand the full nature of light through ratio-empirico-centric method. Thus, in the purview of the above instances, if these two observers argue against "taste", "colour" "wave and light" they will surely be trapped in a paradox. Therefore, in understanding reality holistically, the role of reason and experience must both be complemented at the centre of investigation. As such, a rational centred and empirical centred reality needed to come all together in the formation of a perfect scientific method. This is why the work states that, there is no one way of perceiving reality or carrying out scientific investigations, the critique labelled by both rationalism and empiricism in science may not establish unity in science, as such we state here that both views are necessary factors of grapping the various manifestations of reality to man. The ratio-empirico-centric method is a suitable way to understand, revive and restore these both opposing scientific methods for proper scientific investigations. Without this new model of scientific studies, there will exists a large vacuum in understanding the way nature unfold. In other words, empiricism alone will create a lot of vacuum in science because of the studies of unobservable part of reality (quantum mechanics), and rationalism alone will also bring in a missing link in science because through experience Isaac Newton gave science an absolute laws driven with certainty and deterministic, a role played by his laws in calculus, laws of motion and gravitation.

To solve this crisis of method within the scientific community, non should be seen as superior or inferior to the other or as adulterated, both laws are important in copping the unfolding or dynamic nature of the universe much progress will be achieve if scientist are honest and are imbibing on the culture of ratio-empirico-centric method of science. this can be done through collective commitment and sustenance of good tolerance spirit as demanded by ratio-empirico-centric method.

5.1   Conclusion

The work in philosophy of physics is an ongoing process; it is continuous and has continued to seek for an authentic method of science. The work recommends that the limitations of both methods of science can be overcome from the perspective of ratio-empirico-centric method, because it is seen as the concrete method that can study reality from both the rational manifestations and the empirical manifestations of scientific realities.

It must be noted that, this work suggest that there is no one way method of understanding everything about science, because science is dynamic and the approach should not be static, but be dynamic along the train of postmodernism. The ratio-empirico-centric method as suggested by this work should be one among the many methods that could follow the dynamic nature of science alone the train of scientific developments.

However, we maintain that despite the impacts of both existing methods of science (rationalist and empiricist methods), their successes should not present to us that science could grow at their state of singularity. For there will be a better science when both methods complement to form a unified method of scientific investigation on the nature of the universe through the ratio-empirico-centric method; a philosophical method of scientific studies that integrate the diverse nature of the universe based on the contextual approach that nature manifests to us.

CHAPTER ELEVEN

CRITIQUE OF RATIONALISM IN MODERN SCIENCE

General Criticism of Rationalism

The earliest criticism in the school of rationalism started during the modern philosophical era when British empiricists John Locke, George Berkeley and David Hume took up the task in denying the possibility of knowledge that is aided by reason. Stumpf notes that Locke, Berkeley and Hume moved against the ideas of not only their English predecessors but also the continental rationalists (250) and argue that the foundation of knowledge comes through experience, characterized by our sense of smell, sight, hear and taste. They were great empiricist who took the challenge of investigating the nature of the universe through the purview of experience.

John Locke was born in 1632 at Wrington and also became the founder of empiricism in Britain. Locke explains that the scope of human knowledge is limited to experience as such no man's knowledge can go beyond his experience (Magee 105). He is the author of one of the most popular books in epistemology entitled *An Essay Concerning Human Understanding* published in 1690. Locke believes that when we were born, the mind was like a blank slate (*Tabula Rasa*) in other words, the human mind is more like a blank sheet of paper on which experience writes, and all our knowledge of the universe of science develops from this origin. He contends that all human knowledge are gotten from ideas, these ideas are not rationally based but the ideas that are generated by objects of experience, meaning that all we know must conform to the object of experience. The question critics may raise against Locke's view is; what will happen to invisible objects of reality that cannot be known with the aid of human sensation? Critiques argue that, since Locke's epistemology did not cover the unseen reality, then it has enormous limitations concerning the nature of the universe.

George Berkeley is another unique empiricist, born in Ireland in 1685. He joined the empiricists' movement to pursue a foundation for epistemology. Berkeley begins from experience, as supported by his co-British empiricist, whose main task was to strengthen a knowledge that is gained through sense experience as the foundation of knowledge. Berkeley argues that the foundation of human knowledge begins from experience; as such to be for Berkeley is to be perceived, *esse est percipi* (Stumpf 260), he argues that for the things we do not perceive exist because some other persons perceive them. Thus, all the things we cannot perceive like space, magnitude, and distance also exist because they are suggested to us by experience.

Another empiricist who argued against rationalism is David Hume, born in Edinburgh in 1711. He was social, kind, generous and had a gentle outlook by nature. His work eventually woke Kant from dogmatic slumber and spurred Kant to think afresh about all the problems of epistemology (Derek 117). He was the last and most consistent empiricist among all the British empiricists to argue against rationalism. He argues that, it is only experience that supplies the raw data of experience. Thus, the mind according to Hume can think of countless things like flying horse, golden mountain are all within the construct of experience (Uduigwomen 154). When the mind present the idea of Flying Horse, the mind gains such ideas through experience of a horse and the flying winds from birth that are gotten through perception. Perception for Hume is divided into two; impression and ideas. Impression is the origin of knowledge which we get when we directly perceive an object. It is seen as the origin of our ideas known to be clear, perfect and authentic approach to true knowledge. The question critics may raise against Locke, Berkeley and Hume notions is; what will happen to invisible objects of reality that cannot be known with the aid of human sensation? Critiques argue that, since their epistemology did not cover the unseen reality, then it has enormous limitations concerning the nature of the universe. Thus, we argue that, though empiricism is also an asset to the development of science, but rationalism because of it thought provoking background in explaining reality is the foundation of knowledge, and a key aid to the development of science in modern era.

Micheal Rosen is another scholar who has argued against the enormous ability of rationalism in modern science. In his work entitled *Against Rationalism,* he maintains that though rationalism can be seen as one of the essential aspect of human self development, a full application of its methods weakens man ability to develop practical self-control and self regeneration on *the* perceptive capacity of man. He argues that rationalism to a large extend does not allow man to develop and stimulate the desire for consistent perception which is insufficient for self development. Thus, the effect of this will lead human beings to self-denial in developing self-commands that are established through perception (17). Rosen asked the question, should rationalism be seen as the foundation of human self development? This is what Rosen never agreed, he considered that the essence of perception is to ensure proper application of human sensational power in order to develop the effective use of our sense organs. Rosen conceived knowledge as the inductive science of human nature, and he concluded that humans are creatures more of sensational and practical sentiment than of reason.

Rosen argues further that though rationalism in science should be applied to knowledge acquisition it should not be regarded as a "first order desire" for acquiring a first direct knowledge. Human reasoning ability must be supported by a greater aspect of perception, because rationalism weakens human perceptive ability that develops self-command. Self commands are those active actions that are gotten through sensation, and motivated by self ability to produce a better understanding of the world as directed by perception (Rosen, 18). Conversely, this research argues against Rosen's work and contends that, rationalism is the foundation of modern science because for one to understand the sensible there is a larger application of intelligibility.

The rationalist view of modern science has been criticised by many scholars globally. Firstly, in a work published by Einstein, Podolsky and Rossen (EPR) in 1935 entitled *Can Quantum Explanations of Physical Reality be Considered as Complete?*. In this work Einstein and his colleagues argued against the rationality of quantum mechanics as incomplete explanation of reality (Einstein, et al

41). They did not agree to the position Bohr and Heisenberg claimed about the rationality of quantum mechanics. In reaction to this, Einstein spent a large portion of his career arguing against rationality of quantum mechanics as the last theory of science because for Einstein a theory that can only be explained by mere rationalism without experience or observation cannot be the last theory in science (Laura Ruetsche 201). Einstein and his colleagues conveyed a message that what is appearing as indeterminacy and uncertainty is because we do not know the real nature of reality. Because of this, it appears uncertain and it appears indeterminate, immediately we know the deeper reality, it will reveal that what appears as uncertain or indeterminate is because of our lack of knowledge, because "God does not play Dice" (Wheeler 20). Here the EPR members were still expecting a universe that is created through experience and not rationality. We argue in this work that, the EPR argument though was thought provoking, but the unfolding universe of science is more of rationalism than what the EPR were looking for.

In relativistic physics, the concept of black-hole which was first predicted by general relativity, asserts that there is an infinite hole absolutely black, absorbing light and does not emit out any energy (Craig Callender and Carl Hoefer 173). Thus, in extreme gravity in the black hole, Einstein rational equation seems not to contain; at that level all laws of physics breaks down to infinite; and infinity in mathematics is a number without limits, but in physics it is the collapse of everything about the physical universe. Therefore, there is a problem with Einstein assertion of the black-hole. This lead to what modern physicist called the singularity of the black-hole, meaning scientists can't predict what is going on or the possible outcome of the black-hole. To solve this limitation of the black-hole, quantum mechanics was not enough, meaning there was need to combine gravity to quantum mechanics called quantum gravity (Craig Callender and Carl Hoefer 194). Another problem emerged, the problem of infinite sequence of infinity. This combination failed because quantum mechanics and theory of relativity failed apart, this led to the collapse of physics. A problem yet to be solved even with all tools of rationalism in science, in other words, nature is smarter than what we know through rationalism alone. The methods of rationalism failed to capture quantum gravity; meaning reason alone has enormous limitations.

In view of the above criticism, rationality still has enormous merits in science, especially its fundamental role in theory of relativity and quantum mechanics which are today among the most celebrated theories that have revolutionize science and technology at the modern age.

CHAPTER TWELVE

EVALUATION, SUMMARY AND CONCLUSION

EVALUATION

In this work, we have critically examined the rationalist contents of modern science especially the philosophical approaches to their methods. And from the foregoing, it is clear that science operates or thrives on rationalism more in modern science considering some of the innovative theories that are moving science today like, theory of relativity and quantum mechanics. Which are fundamentally grounded in rationalism, thereby giving prominence to the rational contents of science. We shall further evaluate our argument on rationalism based on the rationality in theory of relativity and rationality in quantum mechanics which are the two innovative theories that shaped modern science.

Rationality in theory of relativity

In 1905 a theory came into the scientific world called the theory of relativity. Theory of relativity is one of Albert Einstein greatest dreams to revolutionized physics and gave support to modern rational content of science. His relativity marks the beginning of modern rationalist moments that came to structure science based on reason. It can be divided into two possible interpretations.

1. The special theory of relativity and
2. The general theory of relativity.

When Einstein was working at the Swiss Patent Office, he had the full time and the energy for some serious evaluative thinking and writing in Physics, especially on serious puzzling problems that relates to the nature of light and motion. In May 1905, at this time he was twenty-six years old, he completed and published three papers, and the last was on *Special Theory Of Relativity* which completely changed the view of the universe (Brennan 57). In the paper Einstein dismissed the problem of absolute motion by denying its existence (Reichenbach 68), he argues that no particular object in the universe is suitable as an absolute frame of reference that is at rest with respect to space. Thus all objects in motion are relative in their frame of reference, and all observers are also relative in their observers' frame of reference (Uduigwomen 19). Einstein's paper is considered by many as one of the greatest work in science that has revolutionized modern thinkers like philosophers and scientists. In the Special Relativity paper, Einstein through the application of rationality predicted that Newton concept of absolute space and time and Maxwell's equations on electromagnetic theory of light could not both be correct (Uduigwomen 123; Russell 17). He held up a lot of rethinking through ratiocination and concluded that one cannot accelerate to the velocity of light and that velocity of light was a constant phenomenon for all observers, regardless of their relative motion, giving us the philosophical implication that absolutism does not give us full descriptions of reality. By further implication as Brennan states, nothing can travel faster than the speed of light, and the speed of light is constant. This idea shattered the dominant Newtonian assumption that an object can travel at unlimited speed as long as enough force is used to accelerate it (58). Einstein questioned Newtonian principles of absolute space and time through his special relativity, and this was possible because he made used of *Thought Experiment*, a kind of experimentation through the mind. He needed no experience or physical observations to establish his theory, but guided by reason, the theory was a success. There are five most important effects of special relativity:

i.   The relativity of simultaneity

ii. Time dilation

iii. Length contraction at speed close to the speed of light

iv. Increase of mass of a fast-moving body and

v.   The relation of mass and energy (Brenan 57).

In the General relativity, Einstein through ratiocination noticed the lapses created by special relativity by not giving adequate explanation of bodies on a state of motion. The general relativity arose through the extension of the principle of relativity to the gravitational force (Gribanov 217). In 1915 Albert Einstein, through ratiocination posited that gravity is different from other known forces, since it is affected by curved nature of space-time (Alozie 83). Because of the distribution of mass and energy, space curved or wrapt and thus, establish a four dimension instead of the three dimensional concept of classical mechanics.

Einstein's general theory of relativity further explains gravity as the curvature of space-time. This concept can be explained by imagining space-time as a rubber sheet on earth were the balls on the rubber sheet bend the sheet around them all the time, somewhat as matter bends space-time in its vicinity (*Encyclopædia Britannica 2014*) this idea is shown below.

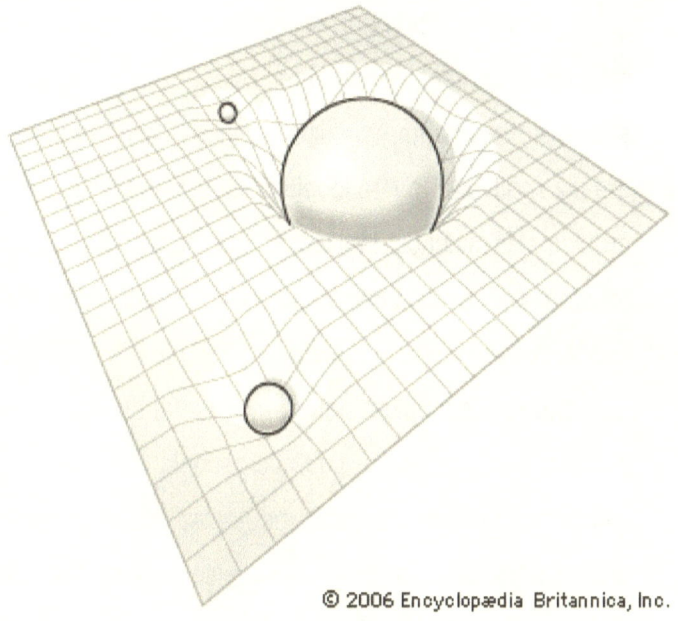

The implication of the above gave scientists a more explicit understanding of the big bang hypothesis, the notion of black-hole and the expansion of the earth surfaces when large objects are placed on earth, something that was never thought of with experience. Thus, it further shows by implication that all moving objects are relative to the observer's frame of acceleration, and all objects moving at the speed of light stop the time (Einstein, *Relativity* 13). Giving us the impression that the speed of light is constant and nothing can move faster than the speed of light, all this predictions were not experienced before prediction; they were only guided by rationality, in what Einstein called the *thought experiment*.

Rationality in Quantum Mechanics

Another moment in science that has proven a remarkable rationalist content in modern science is quantum mechanics. The success of elementary particle physics is another greatest achievements greatly motivated by rationalism in modern science. The fundamental problems in science that could not be accounted for by the laws of Newton and even Einstein relativity could today be explained using quantum mechanics. Quantum mechanics is a branch of physics that give more philosophical significance to the studies of micro particles. It is a theory that studies the sub-microscopic world, such as quarks, atoms, electrons, photons, mesons, neutrons, plasmas and many other sub-atomic particles and their relationship with matter (Uduigwomen 123). These are particles that could not be sensed using any human sensational tools, and as such giving us the impression that, so many things exist outside the one we perceive, and even the ones we perceive are full of limitations. Thus, quantum mechanics has come to build human understanding of the rational world of subatomic particles. In other words, quantum mechanics is rationally based because of its relationship with reason in unfolding the nature of unseen subatomic particles and it origin up till today still becomes a mystery to so many scientists.

The idea behind this theory was first introduced to physics through the work of Max Planck. On December 14, 1900, Max Planck discovered that matter absorbed heat energy and emitted light energy discontinuously or in discrete bits. He later called these *quanta* (Brennan 86). This idea of quanta is what modern physicist calls photons; and photons today can never be seen or heard using human experience or sensation. It brought in, the motivation with great philosophical significance that, nature is unique and consists of particles element which was earlier predicted by the optimistic rationalist Gottfried Leibniz as Monads (William Lawhead 265), but in this sense, Max Planck never called it monads, he called it quarks. Thus, one needs a rational cognitive power to overcome the challenging problems of nature through the study of these quarks that forms the foundation of every reality. Quantum mechanics came in with great renovation concerning the way nature operates, some of these top revolutions came to solve some of the problems of empiricism in science with the aids of rationalism; they include as follows:

1. Quantization of certain physical properties
2. Wave-Particle Duality
3. Quantum entanglement
4. Photoelectric effect
5. The uncertainty and indeterminacy principle, Etc.

The above listed concepts are some of the outcome of quantum mechanics that came to solve some of the problems of science. Quantization of physical properties was first explained in 1900 through Max Planck's discovery of quanta during his lab experimentation on light. Here he discovered that all physical bodies can be quantized when passed through a constant hearted temperature (Hilgevoord 28). Through this discovery, he assert that, every physical object emits electromagnetic radiation with a smooth wavelength spectrum that depends on the temperature of the body in the case of object from every day experience, such as rocks or human body radiation energy. Thus, radiation is in the infrared region of the spectrum where the wavelengths are larger than those of the visible light and is not detectable by human sensation. To the limited human eye it appears that the body is black (Wheeler 9), but in the other way round, light occurs in discrete bundles of energy called quanta.

Quantum entanglement is a core rationalist principle of quantum mechanics, this principle is rationally based, and explains that two or more particles interacts, when their wave-function become entangled no matter the distance, in that some properties of each now depends on what happened to the other particle, (Wheeler 12) distance can never destroy entanglement. Though these particles can be carefully separated, but as far as they are not separated they continue to be entangled. This principle is what has given rise to quantum computing and telecommunication, making it easier to make phone calls from anywhere at any time with no limited duration of time, an innovative principle powered by reason.

In 1905, the same year Einstein introduced special relativity, he explained the photoelectric effect. He postulated that the fundamental quanta of light discovered by Max Planck were actual physical particles. These were later called photons. When a photon of sufficient energy struck a metal, it would kick out electrons; the energy released by the photon body must be greater than the energy binding the electrons of the metals for the electron to be release. According to Einstein the energy emitted out, form the basis for the generation of electric current. Thus, visible photons generally do not have sufficient energy to lift an electron out of the metal, while the photons associated with ultraviolet light have higher energy and can emit electrons (Wheeler 24). Einstein was able to rationally describe the effect using Planck's relationship between energy and frequency with the same constant h (h= Planck constant) to explain a new discovery in science called the Photoelectric effect. It was only possible through the aids of reason that Einstein gave a better explanation of it, as shown in the diagram below.

SCHEMATIC DIAGRAM OF PHOTO-ELECTRIC EFFECT

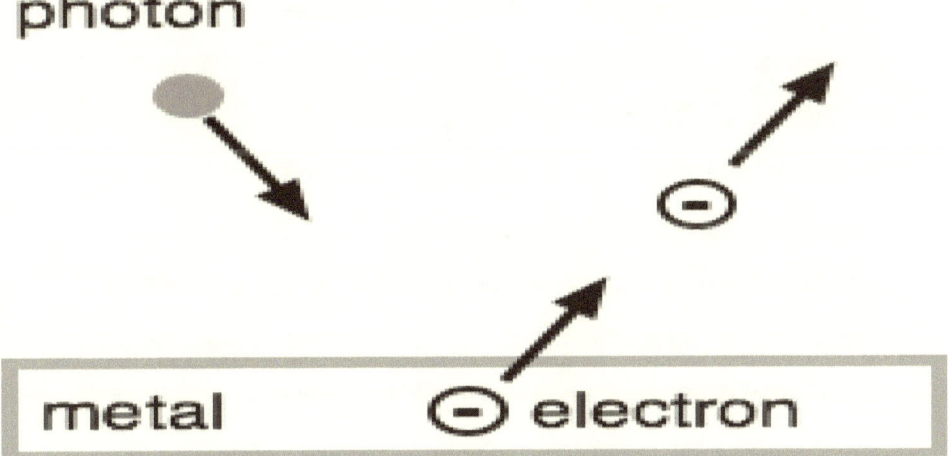

From the above schema, ultraviolet photons kicks up electrons out of metals, where they are relatively loosely bound, which then produce an electric current; this was rationally discovered within the development of quantum mechanics by Einstein in his principle called the photoelectric effect.

On the uncertainty and indeterminacy principles, Neils Bohr's theory had proved that, the position of an electron as well as its speed in a given Bohr orbit could be precisely determined. The question is that, what happens when the electron is on its way to the next orbit? In attempt to answer this question, Werner Heisenberg in 1927 introduced another rational innovation in quantum mechanics known as the "uncertainty principle"; Which says that the momentum and position of a body cannot be simultaneously measured with unlimited precision (Wheeler 14). Except for speed near the speed of light, momentum can usually be approximated as the product of the mass and velocity of a body. However, considering that Newtonian physics has provided us with physical laws of motion that enable physicists to predict the motion of bodies with what seemed, in principle, to be unlimited precision. This implied that the universe itself is one vast machine, a clock work universe in which everything that happens is completely predetermined by what went on before through experience. However such a prediction requires knowledge of both the position and momentum of the body with unlimited precision. Heisenberg refuted this motion showing that an inherent uncertainty exists in the motion of bodies and the best we can do is to predict their average motion. In other words, we cannot predict both the momentum and the position simultaneously, the more precise the speed, the less precise the distance and vice versa. This idea is equivalent to the concept of wave-particle duality or wavicle, were the nature of light acts as wave and also as particle which could not be viewed simultaneously (Alozie 105).

From the above exposition, reason supersedes experience, because of the way nature behaves; our senses do not penetrate into the real nature of subatomic particle universe. Thus, the foundation of science emanates through ratiocination within this understanding. As such, to understand the nature of this kind of universe within the philosophical lime light, there should be no over dependent on empiricism in science, because rationalism gives more explanation about the nature of the universe in modern science.

However, critics may argue that, science should be seen as an integrative cum complementary way of studying the universe from both views of rationalism and empiricism. And some may also argue that rationalism is the first point of call for any scientific knowledge before experience. In other words, knowledge begins with reason and could manifest later with experience depending on the contexts of the scientist. In uncovering the mystery behind nature and how the universe operate within the moments of formation of theory of relativity and quantum mechanics, rationalism through the tools of the mind capture reality first, through this process Einstein was able to predict the theory of relativity which was later confirm empirically years later. Equally within the empiricist tradition raw data that are collected through experience needs reasoning to justify its authenticity. Here, the sense organs may capture images through the eyes and send it to the brain for reasoning to take place. In this case, the work agrees that empiricism is also important in science, but the place of rationalism in giving meanings to fact is the only justifiable way on how to derive truth from nature.

Most scholars content that science operates in such an integrative mood; were the method of science starts from ratiocination stage and later gets confirmation through experience for justification. This approach is essential when dealing with elementary particles within quantum mechanics (Particle Physics) because as particle element they are unobserved, but they exist to the eye through (the formation of particles) when they come together to form matter which could then be seen through experience. Because of this background, this research adopt rationalism in science as a necessary and essential way of understanding the nature, scope and properties of the entire universe; because reality is better understood when reasoning faculty is properly utilized. This work does not state that an absolute one way can better explain the entire reality, thus modern science should not be seen as the handmaid of empiricism alone but rationalism should be seen as the most important aspect of science within the modern era. For this understanding to be sustained; there is greater need for the re-evaluation and revision of scientific literature to promote our understanding of science. Equally rationalism should be seen as the front wheel that coexists with empiricism in science from any contextual point scientist assumes or fines suitable in any stage of scientific investigation.

Summary

The thesis is an attempt to re-assess rationalism in modern science as a method that could explain the progress of modern scientific investigations. Because of the dominance of empiricism in science, the rationalists came up with criticisms using moments in science that are rationally driven and have also given science a new revolution in understanding the mysteries of the universe. Moments of formation of theory of relativity and quantum mechanics were all rationally based, an aspect that investigate reality with enormous success that is today aiding developments in science and technology. The empiricists in modern science believe in the tradition of thought system that supports the aspect of realities that can be accessed through sense data collections and experiment as scientific. On the opposing view, the rationalist also criticised the empiricist that our senses are full of limitations and as such are unreliable for scientist to relent on human senses as the foundation to knowledge. Thus, reason for the rationalist is the only authentic foundation for scientific knowledge.

Concerning the contributions of empiricism in science, the work highlights the various contributions that came into physics (science) as a result of sense data collection and experimentation. In chapter four, we focus attention on the works of Karl Popper, Imre Lakatos, Gottfried Leibniz, and Isaac Newton. The chapter examines these scholars in attempt to build support and to articulate the tenets of rationalism as foundational in the formation of modern science, and whose rational contributions also help in strengthening scientific development within the modern era.

This support has been shown in Isaac Newton's laws of motion and gravitation, which was aided by empirical knowledge and justified by reason. But empiricism as a method of scientific investigation only limits science to the knowledge of the macro world; that is, the world at the very large or the world of matter, the aspect of reality that can only be captured through human experience. It became a very big problem in science, when the method of empiricism (classical mechanics) could no more solve the prevailing problems of physics (science), the problem of the micro world, that is, the world at the very small of invisible particle physics; where classical laws collapsed, and could no more penetrate with its predictive powers.   Thus, the works of scientists and philosophers of science became questionable; paving way for scepticism in science, a big challenge for scientist within the 17th to 19th century quest for knowledge about science. Such as the problem of explanation of the micro world of subatomic particles, quantization of certain physical properties, the problem of understanding wave-particle duality of light (wavicle) the explanation Quantum entanglement, Photoelectric effect and the uncertainty and indeterminacy principle and many others

The work states the various solutions to scientific problems through the aids of rationalism, were reason was rather the only approach to physics (science), as the vital force in solving the problems that could not be counted for using human sense organs. Through the aid of reason, the theory of relativity by Albert Einstein in 1905 was possible, the discovery of quantum mechanics, and the theory of everything (string theory) also followed. All these moments are parts of the major contributions aided by reason as the driving force of scientific development that is today making it possible for us to have a distance communication through wireless mobile phones, tracking of airplanes through Radars and Global Positioning System (GPS) network, the invention of Magnetic Resonance Imaging (MRI) used as brain scanners in modern hospitals are all possible as a result of these developments.

In understanding the ideas and criticisms from these two opposing school, this work establish the impacts of empiricism and rationalism in science and where both differs in terms of methodological approaches. Thus, the work postulates that within the context of empiricism in science, there are indeed enormous limitations and as such, empiricism alone may not be able to propel science to its essential and existential objectives and science should not be seen as an empirical dependent discipline; this is why the work propose rationalism as the foundation of modern science, because its enormous contributions to modern science has help in the growth and development of science and has given a better explanation to the nature of the universe.

The work also expresses circumstances that scientists may not need to neglect empiricist method, because in science today, the laws of Newton are still operational. In other words, classical laws up-till today are still viable to scientific development as captured in chapter three of this work. There are circumstances that scientists also need reason to understand the nature of reality. For example the reality that exist in the realm of subatomic particles. As such there should be no neglected method, all methods of science are important within the context of a given reality.

Conclusion

The work in philosophy of science is an ongoing process. The work recommends that, there should be a revision of the literature in science to reflect the rationalists lining of science. This work suggests that science is now reason-centred within the purview of modern science. And since philosophy is also reason-centred, there should be a review of scientific literature to enable a cross-fertilization of ideas. This will enable philosophers participate in the front role of scientific discourse. By this, the clarion call for philosophers to come back to the discipline is highly recommended.

To enable scientists achieve a better approach when studying the nature of the universe, the work postulates that there should be no empirical dependent justification for science alone. As such the viability of rationalism toward scientific studies should be considered. For this is one of the ways science will grow and be more relevant to philosophers and all lovers of ratiocinative ideas.

This work also suggests that there is no particular or individual method of understanding everything about the universe of science in isolation. Because science is a dynamic discipline and the approach should not be static, but be dynamic along the trend of scientific developments. As such the work posits that despite the impacts of the empirical method of science, its success should not present to us that science could grow in its empirical dependent state without taking cognisance of ratiocination.

REFERENCES

"Aristotle" in www.iep.utm.edu/aristotl/. Retrieved on 9th, June 2015. Web.

"Space-time" *Encyclopædia Britannica. Encyclopædia Britannica Ultimate Reference Suite* Chicago: Encyclopædia Britannica, 2014.

A. d'Abro, *The Rise of the New Physics: its Mathematical and Physical Theories Timerly Tilled Decline of Mechanics* (vol 1) New York: Pover publications Inc; 1939. Print.

Abhay Kumar Singh, *Problems in Physics*, India: Jay Print Pack Limited 2003. Print.

Aigbodioh, Jack A, *Philosophy of Science: Issues and Problems*. Ibadan: Hope Publishers, 1997. Print.

Akira Ishimaru, *Wave Propagation and Scattering in Random media* (vol 1) New York: Harcourt Jovanovich publishers, 1978. Print.

Akpan, Chris O. "An Enquiry into the Influence of Metaphysics on the Development of Modern Science" *Ndunode: Calabar Journal of the Humanities*. Vol 8 no 1, 2009, 109-124. Print.

Akpan, Chris O. "Quantum Mechanics and the Question of Determinism in Science" .*Sophia: and African Journal of Philosophy*, vol.8 no 1, 2005, 72-79. Print.

Albert Messiah, *Quantum Mechanics (Vol 1)*: New York: North Holland publishing Company, 1961. Print.

Albert Messiah, *Quantum Mechanics (Vol 1)*: New York: North Holland publishing Company, 1961. Print.

Allozie, Princewell, *Philosophy of Physics,* Calabar: University of Calabar Press, 2004. Print.

Alozie, Princewill. *Determinism in Biology History and Philosophy of Science* (ed) P. Alozie Calabar: Clearlines, 2001.

Andrew Zimmerman Jones & Daniel Robbins, *String Theory for Dummies* Indiana: Wiley Publishing Inc. 2010. Print.

Andrew Zimmerman Jones and Daniel Robbins, *String Theory for Dummies* Indiana: Wiley Publishing Inc. 2010. Print.

Asouzu I, *The Method and Principles Of Complementary Reflection In And Beyond African Philosophy,* Calabar: University of Calabar Press, 2004. Print

Asouzu, Innocent. *The Method and Principles Of Complementary Reflection In And Beyond African Philosophy,* Calabar: University of Calabar Press, 2004. Print

Ayi A. Ayi. *Understanding Inorganic Chemistry through Problems and Solution,* Calabar: Sues print publishers, 2006. Print.

Barrow John. "Theories of Everything" *Physics and the View of the World* Jan Hilgevoord Ed. Cambridge: Cambridge Press, 1994. Print.

Bette, E.D & Nakanda, E.V. Producing Responsible Citizenship in Nigeria for National Development Through Social Studies Education. American Journal of Social Issues & Humanities Vol. 2 (5) 329-335 2012 Print

Brennan, Richard P. *Heisenberg Probably Slept Here: The Lives Times, and Ideas of the Great Physicists of the 20th Century.* Canada: John Wiley & Sons, inc. 1997. Print.

Carl Craver, "Structure of Scientific Theory" *Blackwell Guides to the Philosophy of Science.* Peter Machamer and Michad Silberstein Eds. Chicago: Blackwell Publishers Ltd, 2002. Print.

Chalmers A. F. *What is this Thing Called Science?: An Assessment of the Nature and Status of Science and its Methods* Buckingam: Open University Press, 1990. Print.

Corkcroft & Walton "E=Mc^2" in http://zebu.uoregon.edu/2004/hum399/lec18.html Retrieved 1/12/2016

Craig Callender and Carl Hoefer, "Philosophy of Space-Time Physics" in *Blackwell Guides to the Philosophy of Science.* Peter Machamer and Michad Silberstein Eds. USA: Blackwell Publishers Ltd, 2002. Print.

Daniel Kleppner and Robert J. Kolenkow, *An Introduction to Mechanics* Singapore: McGraw-Hill, Luc. 1973. Print.

Derek, Johnson, *A Brief History of Philosophy: From Socrates to Derrida.* New York: MPG Books Ltd. 2006

Edet, Mesembe, "Being as a Missing Links" on *Journal of Complementary Reflection: Studies in Asouzu* Vol. 1/1 ISSN 2026-6545 October 2011, 28-32. Print.

Enu, D. B., Kalu, I. M. & Obi, Florence B. (2006). Secondary School Students' Perception of Societal Value for Paper Qualification and Cheating Tendency in Examinations. *Journal of the faculty of Educational Studies, 170-176, Ghana.*

Olah, S., Undie, J. A. & Enu, D. B. (2009). Globalization and the realities of Cross Border Higher Education in an Unequal world. *International Journal of Higher Education Research (IJHER).68-81, Ghana*

Eneji, C. V. O., Qi Gubo, Q., Umoren, G. U, Omoogun, A. C., Oden, S. N, Enu, D. B. & Edet, P. B (2009). Socio-economic impacts of the Cross River National Park, Nigeria. *Journal of Agriculture, Biotechnology and Ecology, 2 (1)57-58, China.*

Enu, D. B., Oru, P. B., Omoogun, A. C., Oden, S. N. & Domike, G. (2009). A survey of Bush Meat Extraction within the Forest Communities of Ojok and Ekonganaku in Akamkpa in Respect of Environmental Curriculum in Education. *Journal of Agriculture, Biotechnology and Ecology, 2(1) 78-85, China.*

Enu, D. B., & Esu, A. E. O (2011) Re-Engineering Values Education in Nigerian Schools as Catalyst for National Development. *International Education Studies 4, (1)147-153, Canada.*

Enu, D. B. & Ugwu, U. (2011) Human Security and Sustainable Peace Building in Nigeria: The Niger Delta Perspective. *Journal of Sustainable Development 4,(1), 254-259, Canada.*

Enu, D. B. (2012), Enhancing the Entrepreneurship Education in Nigeria. *American Journal of Social Issues & Humanities, 2(4), 232-239, USA.*

Enu, D. B. & Effiom, Veronica Nakanda (2012) Producing Responsible Citizenship in Nigeria for National Development through Social Studies. *American Journal of Social Issues & Humanities; 2(5), 329-343, USA.*

Enu, D.B. (2014), Predictive Influence of Academic Self-concept and Motivational Arousals in examination Cheating Tendency among Students in Cross River State - Nigeria; *British Journal of*

Education, Society & Behavioural Science; 4(3), 383-391, Britain.

Enu, D. B. & Eba, Maxwell Borjor (2014) Teaching for Democracy in Nigeria: A Paradigm Shift. *Journal of Higher Education Studies;* 4 (3), Canada.

Enu, D. B. (2015). Teaching for social justice: An exploration of ethnic discontent in Nigeria. Accepted for publication in *Journal of Education and Practice, USA.*

Enu, D. B. & Unimke, S. A. (2015). Exploring current trends of Social Studies curriculum in Nigeria within the context of global competiveness. Accepted for publication in *International Journal of Humanities and social Science, USA*

Enu, D. B. (2016). Social Educators' Contribution to Educating for Peace and Security in Nigeria. *Accepted for publication in Journal of Education and Practice, USA.*

Enu, D. B., Opoh, F. A. & Esu, AEO. (2016). Evaluation of Cross River State access to matching grants for the implementation of UBE policies between 2010 and 2014. Accepted for publication by *Journal of Education and Practice. USA*

Enu, D. B. (1999) Environmental Education as a Major Innovation in the Nigerian School Curriculum. *West African Journal of Educational Research 2(2). 120- 126.*

Enu, D. B. (2000). Educational Accountability: The basis for a Realistic Education in the 21st Century. *West Africa Journal of Educational Research; 3(1), 73–76.*

Enu, D. B. (2000). Actualizing the UBE curricula through Community Participation. *International Journal of Research in Basic and Life-Long Education. 1(1&2), 52-59.*

Enu, D. B. & Dibang, F. E. (2001). Facilitating the UBE Curricula Implementation through Sustainable Human Resources Development. *West African Journal of Research and Development in Education; 8(1), 61-65*

Enu, D. B. (2002). Curriculum Approach of Mobilization of the Rural Communities for Sustainable Development Planning in Nigeria. *Global Journal of Educational Research; 1(2), 43-48.*

Enu, D. B. (2004) Evolving a Functional Human Rights Education Curriculum through Sustainable Human Resources Development. *Nigeria Journal of Curriculum Studies; 11(2), 18-24.*

Enu, D. B. (2005). New Challenges for Social Studies Curriculum Content in a Globalised Social System. *Journal of World Council for Curriculum and Instruction; 5 (1 or 2).*

Enu, D.B. (2006). Attaining Environmental Sustainability through the Formal School Curriculum. *Journal of Curriculum Organization of Nigeria; 2 (1), 101-106.*

Enu, D. B., Omoogun, A. C., Oji, G. O., Ekuri, E. E., Okeme, I. (2008). Achieving Quality Assurance in Nigeria University System through Strategic Human Resources Development. *International Journal of Educational Research; 4 (1), 161-166.*

Enu, D. B., Esu, A. E. O., & Kalu, I. M. (2008). School Environmental Variables As Predictors of JSS3 Students Social Studies academic achievement Cross River State. *Nigerian Journal of Curriculum Studies; 15 (2), 85- 95.*

Omoogun, A.C. & Enu, D. B. (2008). Quality and Quality Assurance: The Imperative for Improved Teacher Quality in Nigerian. *Journal of Curriculum Studies; 15 (4).*

Enu, D B. & Omoogun, A. C. (2009). Can the School Curriculum Teach students to be Peace Makers in a Multi-ethnic Society? *Journal of Curriculum Studies; 19, (4), 125-134.*

Omoogun, A C & Enu, D. B. (2009). The challenges of the University as Agent of Social Change: *Journal of Curriculum Studies, 16, 258-266.*

Esu, A. E. O. & Enu, D. B. (2009). Values and Ethics in School Curriculum. In U M O Ivowi, Nwufo, K., Nwagbra, C., Ukwungwu, J., Emah, I. E. & Uya, G. (Eds). In *Curriculum Theory and Practice*: Curriculum Organization of Nigeria. 284-290

Enu, D B & Omoogun, A C & Okeme, I. (2009). University Education Curriculum and Emerging Challenges of .Responding to Global Trend. *Nigerian Journal of Curriculum Studies; 3, (1&2), 50-58.*

Omoogun, A C, Enu, D. B. & Akpan, D. S. (2009). Ensuring Sustainability of Universal Basic Education: The Four Wheel Dimensions. *Nigerian Journal of Curriculum Studies; 3,(1&2), 229-238*

Enu, D. B. (2005). *Teaching Environmental Education in Formal and Informal Settings*. Calabar: Wusen Publishers.

Egbai, Uti. "Is Quantum Mechanics a Complete Theory?: a Philosophical Defence of Einstein's Position" in *Sophia: and African Journal of Philosophy*, vol.8 no 2, 2006, 14-19. Print.

Egbai, Uti. "The Emergence of Subjectivism in Physics and the Possible Implications" in Sophia: *and African Journal of Philosophy and Public Affairs*, vol.10 no 1, 2010, 104-109. Print.

Einstein, A. "Physics, Philosophy and Scientific Progress". *The Journal of the International College of Surgeon's,* 1950, quoted in D.P. Gribanow Einstein Philosophical Views and the Theory of relativity. 200-204. Print.

Einstein, A., Podolsky, B. And Rosen, N. "Can Quantum Mechanical Description of Physical Reality be Considered Complete?" in *Physical Review. 1935. 41-47. Print.*

Einstein, Albert. *Out of My Later Year*, 5th ed. New York: McGraw Hill, 1984. Print.

Enu, D. B., & Esu, A. E. Re-Engineering Values Education in Nigerian Schools as catalyst for National Development. International Education Studies. 2011. 147- 153. Print

Ephraim Chiedozie "The Early Modern Science of Copernicus and Kepler" in *Philosophy and the Rise of Modern Science* A. F. Uduigwomen Ed Uyo: El-John Publishers, 2011. Print

Essien, Ephraim. "Einstein's Relativity Theory and the Structure of the Universe" *Sophia:An African Journal of Philosophy and Public Affairs*, vol.10 No 1, 2007. 220-228. Print.

Fagothey, Austin. *Right and Reason: Ethics in Theory and Practice.* California: Mosby Company, 1976

Feyerabend, Paul, *Science without Experience* in *The Journal of Philosophy*, vol. 66 n 22, 1969, 791-795. Print.

Feyerabend, Paul. *Against Method; Outline of an Anarchistic Theory of Knowledge.* London: New Left Books, 1975. Print.

G.W.F. Hegel, *Natural Law: The Scientific Ways of Treating Natural Law, Its Place in Moral Philosophy, and Its Relation to the Positive Sciences of Law.* Philadelphia: University of Pennsylvania Press, 1975.

Glenn Borchardt , The Physical Meaning of E=mc2 in www. http://www.scientificphilosophy.com/Downloads/The%20Physical%20Meaning20of%20E=mc2.pdf Retrieved 12/10/16

Gribanov D.P. *Albert Einstein's Philosophical Views and the Theory of Relativity*, Moscow: Progress Publishers, 1987. Print.

Gribanov D.P. *Albert Einstein's Philosophical views and the theory of Relativity*, Moscow: Progress Publishers, 1987.

Hans C. Ohanian, Einstein's E=mc2 Mistakes in http://arxiv.org/ftp/arxiv/papers/0805/0805.1400.pdf

Hawking, S.W. *A Brief History of Time.* London: Bantam Book, 1988. Print

Heisenberg, Werner. *The Representation of Nature in Contemporary Physics.* London: Unwin Hyman Publishers, 1955. Print

Herbert Butterfield, *The Origin of Modern Science.* New York: The Free Press, 1966. Print.

Hermann Bondi, *Relativity and Common Sense: A New Approach to Einstein* Great Britain: Heinemann books, 1968. Print.

Igwe Dennis E. "Newton and the Theory of Universal Gravitation" in *Philosophy and the Rise of Modern Science* A. F. Uduigwomen Ed Uyo: El-John Publishers, 2011. Print

Ijiomah Chris *"The Crisis in Geometry and the Rise of Relativistic Logic in Twentieth Century"*, Quoted in *Global Journal of Humanities Vol.* 4 No 1&2, 2005. Pp 5-9.

Ijiomah Chris O. *Harmonious Monism: a Philosophical Logic of Explanation for Ontological Issues in Supernaturalism in African Thought.* Calabar: Jochrisam Publshers, 2014. Print.

Ijiomah Chris, *Modern Logic: A Systematic Approach to the Study of Logic.* Owerri: A.P Publishers, 1995.
Index: Google Scholar (Nigeria)

Iroegbu Panteleon, *Metaphysics: The Kpim of Philosophy* Owerri: International Universities Press Ltd., 1995

Jan Hilgevoord. *Physics and our View of the World,* New York: Cambridge University Press, 1994. Print.

Jerry B. Marion, *The Universe of Physics: A Book of Reading.* New York: John widely and Son, Inc. 1970. Print.

Jim Woodward "Explanation" in *Blackwell Guides to the Philosophy of Science.* Peter Machamer and Michad Silberstein Eds. Chicago: Blackwell Publishers Ltd, 2002. Print.

John Archibald Wheeler. *The Spooky Quantum* in www.google .com/search/index/html. Web. 14/4/2015.

John Cottingham Ed, *The Cambridge Companion to Descartes* New York: Cambridge University Press, 1998. Print.

John Fauvel (et al) *Let Newton Be:* New York: Oxford University press 1988. Print.

John R. Reitz and Frederick J. Milford, Foundation of Electromagnetic Theory, England: Addison-Wesley Publishing Company, Inc. 1960. Print.

John Stuart Mill, *System of Logic Ratiocinative and Inductive: Being A Connective View Of the Principles of Evidence and the Methods of Scientific Investigation.* New York: Longman Press, 1898. Print.

Kant, Immanuel. *Critique of Pure Reason.* Trans Max Muller. New York: Doubleday and Company, 1966. Print.

Kilmister, C. W. *Special Theory of Relativity* New York: Pergamon Press 1970. Print.

Lawhead, William, *The Voyage of Discovery: A Historical Introduction to Philosophy.* Australia: Wadsworth Group, 2002. Print.

Magee, Bryan. *The Story of Philosophy* Spain: Dorling Kindersley, 2001.

Mamadu, Terver Titus. "The Experimental Method of Gilbert and Bacon" in *Philosophy and the Rise of Modern Science* A. F. Uduigwomen Ed Uyo: El-John Publishers, 2011. Print

Marion, Jerry B. *Physics in the Modern World* London: Academic Press, 1981. Print

Mendie, Patrick Johnson. "The End of Road Thesis in Science" in *Contemporary Journal of Arts and Science,* Vol. 1.1 2015, 120-131. Index: Google Scholar (Netherland)

Mendie, Patrick Johnson. "Asouzu's Critique of Philosophy of Essence and Its Implication for the Growth of Science" in *Philosophy*

Study, Vol. 5 No: 5, 2015. 233-243. Index: Google Scholar (China)

Mendie, Patrick Johnson. "Multiculturalism and Political Development In Nigeria: In the Purview of Integrative Humanism" in Journal of *Integrative Humanism,* Vol. 5 No: 1, 2015. 143-154.

Index: Google Scholar (Ghana)

Mendie, Patrick Johnson and Emmanuel Eyo. "Environmental Challenges And Axiology: Towards A Complementary Studies In Eco-Philosophy" in Journal of *Integrative Humanism,* Vol. 7 No: 1, 2016. 144-150. Index: Google Scholar (Ghana)

Mendie, Patrick Johnson and Joseph Essien. "String Theory: A Realism Or Idealism" in Journal of *Integrative Humanism,* Vol. 7 No: 1, 2016. 151-158. Index: Google Scholar (Ghana)

Mendie, Patrick Johnson. "The Confluence of Philosophy and Biology: An Excavation of Philosophical Issues in Molecular and Developmental Biology" in Online Journal of Health Ethics, Vol. 12 No: 2, 2016. Index: Google Scholar (United State of America)

Mendie, Patrick Johnson "The Future of Metaphysics in African Philosophy: an Ika-Annang perspective" in *American Journal of Social Issues and Humanities.* Vol. 5 No: 1, 2015, 373-383.

Index: Google Scholar (United State of America)

Mendie, Patrick Johnson (Etal) "An Examination of the Nexus between Thomas Hobbes' Concept of Human Nature and the theory of the State" in *SOPHIA: An African Journal of Philosophy and Public Affairs.* Vol. 16 No: 1, 2015, 101-107.

Index: Google Scholar (Nigeria)

Mendie, Patrick Johnson "Pricing the Environment and the Question of Global Environmental Network (GEN) A Philosophical Approach" in *NDUNODE: Calabar Journal of Humanities.* Vol. 12 No: 1, 2017, 362-368.

Index: Google Scholar (United State of America)

Mendie, Patrick Johnson (et al) "Logical Positivism versus Thomas Kuhn" in *THE LEAJON: An Academic Journal of Interdisciplinary Studies.* Vol.6 No. 1, 2014. 197-216 (Nigeria)

Mendie, Patrick Johnson "The Role Of Doxastic and Non-Doxastic Theories: An Epistemic Appraisal" in G.O. Ozumba (Ed) *The Mirror of Philosophy.* Calabar: El-John Press, 2014 (Nigeria)

Mendie, Patrick Johnson "Isaac Newton" in *The A. F Uduigwomen (Ed) A Critical History Of Philosophy (Vol.2). Calabar: Ultimate Index Press, 2016 (Nigeria)*

Mendie, Patrick Johnson "Robert Boyle" in. *A. F Uduigwomen (Ed) A Critical History Of Philosophy (Vol. 2). Calabar: Ultimate Index Press, 2016 (Nigeria)*

Mendie, Patrick Johnson "Postmodernism and Science" in. *G. O. Ozumba et tal (Eds) Critical Essays on Postmodernism. United Kingdom: Lulu Press Inc, 2017. 257-268 (United Kingdom)*

Mendie, Patrick Johnson "Postmodernism and the Theory of Relativity" in. *G. O. Ozumba et tal (Eds) Critical Essays on Postmodernism. United Kingdom: Lulu Press Inc, 2017. 269-275 (United Kingdom)*

Mendie, Patrick Johnson "Postmodernism and the Post-Modernization of Logic" in. *G. O. Ozumba et tal (Eds) Critical Essays on Postmodernism. United Kingdom: Lulu Press Inc, 2017. 90-98 (United Kingdom)*

Michael Rosen, "Against Rationalism" http://scholar.harvard.edu/files/michaelrosen/files/against_rationalism.pdf (30/08/2015)

Nathan Spielberg and Bryon D. Anderson, *Seven ideas That Shocks the Universe,* Canada: Courier Stoughton, 1985. Print.

Newton-Smith W.H. *The Rationality of Science* London: Routledge Press, 1981

Nielson J. Rud Ed. *Neils Bohr Collected Works (volume 4): The Periodic System* New York: North-Holland publishing company 1977. Print.

Nielson Rud Ed. *Neils Bohr Collected Works (volume 4): The Periodic System* New York: North- Holland publishing company, 1977. Print.

Ogbozo, Chrysanthus Nnaemeka. *Philosophy of Science: Historical and Thematic Introduction.* Enugu: Claretian Communication, 2014

Ozumba G. O, *A Concise Introduction to Epistemology,* Calabar: Jochrisam Publishers, 2001. Print.

Ozumba, G. O. "ISMS in Philosophy". *A Concise Introduction to Philosophy and Logic,* Uduigwomen, A.F. and Ozumba G.O. (Eds.), Calabar: Jochrisam Publishers, 1995. Print

Pagels, Heinz. *The Cosmic Code: Quantum Physics as the Language of Nature.* Toronto: Bantam Books, 1982. Print.

Peter Woit "String Theory and Post-Empiricism" in https://scientiasalon.wordpress.com/2014/07/10/string-theory-and-the-no-alternatives-argument . Web. 15/6/2015.

Planck, Max. *"The origin and Development of quantum theory"* in *Great ideas in modern science* Robert W. Marks (Ed.) Toronto: Bantam Books, 1967 Print.

Popper, Karl. *The Logic of Scientific Discovery* London: Hutchinson, 1934.

Popper, Paul. *Conjectures and Refutations: The Growth of Scientific Knowledge.* London: Routledge, 1989

Richard P. Brennan. *Heisenberg Probably Slept Here,* USA: John Wiley & Sons Publishers, 1997. Print.

Richard P. Feynman, *The Feynman, Lectures on Physics: Quantum mechanics* New York: Addison-Wesley publishing company 1965. Print.

Robert A. Sungenis, Ph.D. E = MC^2 — Not Einstein's Invention http://galileowaswrong.com/wp-content/uploads/2013/06/E-equals-mc2-Not-Einsteins-Invention.pdf

Roberto Torrehi, *The Evolution of Modern Philosophy: The Philosophy of Physics* U.S.A. Cambridge university press, 1999. Print.

Roberto Torrehi, *The Philosophy of Physics*, USA: Cambridge University Press, 1999. Print.

Ruetsche, Laura. "Interpreting Quantum Theories" in *Blackwell Guides to the Philosophy of Science* Peter Machamer and Michad Silberstein Eds. USA: Blackwell Publishers Ltd, 2002. Print.

Serway, Raymond (et al). *Principles of Physics.* San Diego: Harcourt Brace College Publishers, 1998. Print.

Sharmu I. K. *A Textbook of Physical Chemistry.* Delhi: city printers, 2009. Print.

Shashi Chawla. *A Text Book of Engineering Chemistry*, Delhi: Educational and Technical Publishers, 2010. Print.

Shirkor D. V. (et al), *Introduction to the Theory of Quantized Fields vol III)* New York: John Wiley and Son, 1959. Print.

Thouless, D. J. *The Quantum Mechanics of Many-Body Systems* London: Academic press 1961. Print.

Uduigwomen, A. F, A *Textbook of History and Philosophy of Science*, Aba: AAU Vitalis Book Company, 2007. Print.

w.w.wikipedia.com/search/quantum,mechanics/index/html/11/Oct.2016.

Walter Greiner & Bernat Muller, *Quantum mechanics symmetries* Berlin: Springer, 1989. Print.

Wong Chee Leong & Yap Kueh Chin, *Conceptual Development of Einstein's Mass-Energy Relationship* Cited in http://files.eric.ed.gov/fulltext/EJ848449.pdf Retrieved 12/10/16

Zukav, Gary. *The Dancing Wuli Masters: An Overview of the New Physics.* London: Rider and Company, 1979.